T0141701

REFLECTIONS: A MEMOIR

Distinguished Visiting Scientist
NASA Goddard Space Flight Center
Greenbelt, Maryland 20771

REFLECTIONS: A MEMOIR

by

David Atlas

© Copyright 2001 by the American Meteorological Society.
Permission to use figures and brief excerpts from this
monograph in scientific and educational works is hereby
granted provided the source is acknowledged. All rights
reserved. No part of this publication may be reproduced,
stored in a retrieval system, or transmitted, in any form
or by any means, electronic, mechanical, photocopying,
recording, or otherwise, without the prior written
permission of the publisher.

ISBN 1-878220-46-2

Published by the American Meteorological Society
45 Beacon Street, Boston, MA 02108

Ronald D. McPherson, Executive Director
Keith L. Seitter, Deputy Executive Director
Kenneth F. Heideman, Director of Publications
Kate O'Halloran, Copy Editor

Printed in the United States of America
by Sheridan Books

Table of Contents

Dedication

This memoir is dedicated to my colleagues who stimulated me and filled the gaps in my scientific arsenal, and to my wife Lucille, with love; she provided my emotional balance wheel through highs and lows.

Chapter 1

A Memoir—Why?

It was not until this memoir was well underway that I fully appreciated why I had to do it. As the remaining years dwindled away, I got a yearning to take stock of the 56 years that I have spent as a meteorologist. I needed to take a retrospective look while my memory still served me, with the perspective of one who has seen the evolution of the discipline in general and of radar meteorology in particular. From this view at the apex of the remarkable scientific developments of the second half of the twentieth century, I wondered how the work that my colleagues and I had done in a seemingly helter-skelter fashion fitted into a coherent picture of the present state of the art. I also thought it would be fun to relive the exquisite joys I have experienced as a researcher and to reevaluate some of the painful periods.

At first, I approached this task reluctantly, because I wanted it to be a story about the scientific process rather than an account of my personal activities, and it was not clear how this could be done. Finally, I realized that my own story was the foreground for a bigger picture that comprised the milieu in which we worked. My hesitancy was also overcome by encouragement from a number of my colleagues to put my recollections to paper for the record. I am doing so because I am one of the last of the surviving pioneers who is still active, and whose career began with the birth of radar during WWII. In a way, I serve as the corporate memory of that wondrous group of people who have left the scene. I hope I do them justice.

An outline of the evolution of radar meteorology can be gleaned from the literature, but nowhere can one find the important details— the nuances, the role of chance and opportunity, and the human dimensions. Some of the men and women involved were my colleagues and mentors; some were the innovators whose work stimulated my own; we were all competitors in this exciting race; and most were my cherished friends. I knew most of them, whether in the United States or abroad. I have also participated in a variety of tangential disciplines such as severe storms, precipitation physics, weather modification, oceanography, and aviation. I have worked both in academe and in various government institutions. And I have played a modest role in public policy. The result is that I have a broad perspective that may prove useful to those following in our footsteps. Finally, I shall take the liberty to provide some comments on the conduct and nature of the scientific process and the institutions that manage it.

This is not a history of the field of radar meteorology, such as the excellent accounts that have been written by Hitschfeld (1), Rogers and Smith (2), or the series that appear in the book *Radar in Meteorology* (3). Rather, it is a personal story of the various people and institutions with which I was associated, and the most significant developments in which we participated. I have emphasized those endeavors that either were pivotal, were particularly great fun, or offer lessons good or bad, and in which I have played a significant role. For me this has been a great adventure story.

Of course, this journey started at an opportune time and could not have been accomplished alone. I was fortunate to be there at the birth of radar meteorology. This was a time when the total weather radar community comprised some 50 people who knew each other. We were close-knit and met frequently in small conferences where we delighted in telling each other about our work and were not reluctant to criticize. New insights and discoveries came fast and furiously for there was something novel to see and explain every time we turned on the radar. Our friendships blossomed into those mysterious symbiotic collaborations that yield explosions of discovery. I have had the good fortune to have more than my share of such relationships; they are an intrinsic element of this story.

I've attempted to avoid excessive scientific and technical detail to permit understanding by the lay reader. For those who are interested, you will find a brief tutorial on some of the fundamentals of radar meteorology in Appendix C. If a particular paragraph is incomprehensible, I suggest skipping it. Numbers in parentheses refer to references found at the back.

CHAPTER 2

Early Years

I am the third child born to Rose (Jaffee) Atlas and Isadore Atlas, immigrants from Russia and Poland, respectively. My mother, one of six children, worked at a sewing machine in a New York factory from the age of 12 and never finished grade school. My father came to the United States with an uncle in 1912 at the age of 16, worked in a ladies' hat factory, and finished high school at night. Except for his sister Esther, who immigrated to Mexico City in 1928 and made contact with him in 1942, my father never again saw his parents or siblings. My parents were married in 1917. My brother Moe was born in 1918, my sister Mae was born two years later, and I came along in 1924.

My family was of less than modest means, but except for the movies, I was not aware of it. There was always sufficient food on the table. With four surviving aunts and their families on my mother's side, my grandfather, his sister, her five sons and a daughter, and their offspring, we formed a close-knit Atlas-Jaffee clan centered mostly in the East New York section of Brooklyn. I miss those old days and family gatherings. We now yearn to see our grandchildren more often than two or three times per year.

The decade of the 1930s spanned my grade- and high-school years. Unlike my sister and brother, I did well in school, revered my teachers, and was motivated by them. My parents also took great pride in my achievements and showered their "baby" with love. From the ages of 13 to 17, I also played the accordion enthusiastically and hoped to be a virtuoso until I recognized that I had no natural talent for it. I abandoned the accordion in an emotional encounter with my teacher.

Having gone through "rapid advance" in junior high, I finished the ninth grade a year early. I then commuted to Boys High School because of its high reputation, rather than attending the local school. I enjoyed those years greatly, got good grades, edited the Spanish magazine, and presided over the Pan American club. I also worked in the school library.

I graduated in January 1941 at the age of 16 and decided to postpone college until September, since I felt I was too young. Instead,

I worked in the main New York Public Library (42nd St. and 5th Ave.) for $11 a week. With my name, they assigned me appropriately to the map room. Later, I worked in the patent department, where I used my spare time to search for novel and "Rube Goldberg" inventions. I also took a night class in typing.

I started to attend the College of City of New York (CCNY, now University) in September 1941, commuting three hours per day on the subway and doing most of my homework on the train. My special joy was freshman physics under Prof. Zemansky. He opened my eyes to science and made it fun. My aim was electrical engineering. With the outbreak of war on December 7, 1941, I felt the need to accelerate my education and so took an entire year of calculus during the summer of 1942. Having turned 18 in May (with elementary knowledge of the equation for a tuned radio circuit), I was able to get a job at Western Electric in Kearny, New Jersey. I worked on the production line for aircraft radios. My summer routine was: up at 6 A.M., CCNY by 8 A.M., classes till noon, homework till 2 P.M. and the ride to Kearny for work starting at 3 P.M. I worked until 11 P.M. or later, and got home by 1 or 2 A.M. to repeat the cycle. The reward was the munificent salary of $35 per week, a figure that astonished my family because my father was only earning about $50 per week. My first job on the production line was tracking aircraft condensers—I was the only male on the line. But I was much too slow because every time I had to adjust the end condenser plate I went through the reasoning process: increase the plate spacing to decrease the capacitance and increase the tuned frequency. The ladies, who simply watched the needle on the capacitance bridge, were doing 60 or more condensers during a shift, while I was pleased to complete 10. But the foreman recognized my "superior" talent and switched me to trouble-shooting aircraft radios that had been rejected. The summer routine was so exhausting that I was eager to return to the less daunting schedule of school and library work in September.

In the fall of 1942, I found a newspaper item announcing openings for recruits into premeteorology training for the Army Air Corps. (I must admit that I did not know exactly what meteorology entailed.) Anticipating the possibility of being drafted into the service, but unable to find any specific information at the recruiting offices in New York, I called Prof. Joseph Kaplan, the program coordinator at the University of Chicago. Thus, I was among the first in the city to get an application. To my chagrin, I was rejected. But this could not have been right because my friends with lesser grades were accepted, so I discarded the rejection and applied again; that time I was accepted. Call it naiveté; call it persistence; it worked.

I enlisted in December 1942, and reported for active duty at the Bronx campus of New York University on March 8, 1943. On the subway a young man noticed that the name inscribed on my bag matched one of those listed on his orders to active duty. He introduced himself as Louis Battan, who was to become my roommate, lifelong friend, and colleague. Five years later Lou was best man at my wedding (Plate 2). A tribute to Lou Battan and an account of our stay at NYU appears in *Radar in Meteorology* (3).

We spent six months as enlisted men coming up to speed on the prerequisites for meteorology training, and nine months as aviation cadets studying meteorology. Among our instructors were Hans Panofsky (dynamics), Yale Mintz (climatology), and lieutenants Bill Gordon and Bob Fleagle (synoptic meteorology). The four were to become close friends and colleagues. Years later, Yale joined my laboratory at Goddard Space Flight Center after an eminent career at UCLA. I learned most about turbulence from Hans' book on the subject (with Lumley) and from his amusing lectures at conferences. And Gordon and Fleagle were thoughtful, highly respected advisors and sometime fellow committee members under the American Meteorological Society, the University Corporation for Atmospheric Research (UCAR), and the National Research Council (NRC). Meteorology training was so intense that it could not be called enjoyable, but we absorbed all we could. I was never sure of what I knew, so it was a surprise when I graduated first in the class. Lou Battan was also in the top 10.

We were commissioned second lieutenants on June 5, 1944, the day before D-Day in Europe. I was assigned as a forecaster at a flying school at Courtland, Alabama, where I remained for only three weeks before being reassigned to the Weather Instrument Training School at Seagirt, New Jersey. There we studied basic electronics for two months before proceeding to Harvard/MIT for radar school. I recall with embarrassment the vacuum tubes I blew when I connected the 110-volt power to the 6-volt filaments.

Radar training at Harvard and MIT and its influence on the field is described briefly by Rogers and Smith (2). The actual experience was a test of fortitude. The first four months were spent at Harvard, where we studied basic electronics, radio, and electromagnetic theory. But none of us had any idea that we were studying radar (the very word was classified) until we moved over to the MIT Harbor Building in downtown Boston where we actually got our hands on radar components and handbooks. There we worked from early morning to late at night, even doing our homework under lock and key since our notebooks were classified. It was a daunting experience. I greatly enjoyed working with hardware and tracking aircraft with a rooftop radar. When we finished class at noon each Saturday I rushed to South Station to catch the train to New York, where I spent a total of 20 hours visiting my family and friends before catching the train back to Boston by 10 P.M. on Sunday.

I petitioned NYU for my bachelor's degree on the basis of all the credits accumulated there, at CCNY, and Harvard and MIT. The B.Sc. was awarded in February 1946. I was discharged from the Army in October 1946. After a period at Ohio State University, I returned to graduate school at MIT in 1948, where I received my M.Sc. in 1951 and D.Sc. in 1955 while I was working at the Air Force Cambridge Research Laboratories (AFCRL). That too was a hectic experience because I could only take three hours per week of classes, and we were raising a family. But I was grateful to the U.S. Army for the GI bill, without which I could never have continued my education. Surely, the GI bill profoundly influenced the course of the country for all time by permitting students from the lower economic strata to complete college and graduate training.

My curriculum vitae is presented in Appendix B.

CHAPTER 3

All Weather Flying Division, 1945–1948

Upon graduation from radar school in April 1945, I was assigned to the newly formed All Weather Flying Division (AWFD) at Wright Field, Dayton, Ohio to do research and development (R&D) on weather radar for flight safety. The rest of my class was assigned to set up APQ-13 (airborne) radars to be used as ground-based systems for storm detection at various Air Corps bases. (The radar names were symbolic of their function: A = Airborne, P = Pulse, and Q = Special purpose, in this case, bombing.) The prior class had been assigned to fly B-29 weather reconnaissance out of Guam and Tinian in the Pacific on long missions to Japan. My assignment was a stroke of luck for it put me into a position where I was to get an early taste of research and to come into close contact with the scientific community.

The day I arrived, my commanding officer, Lt. Col. Philpott, suggested that we make a training flight to Kansas City (where he had a lady friend) to familiarize me with the APS-15 navigation radar. I convinced him to delay a day to allow time for me to read the handbook and try it out on the ground. When we took off the next day I was pleased with my ability to get the radar turned on and working. We were flying above an altocumulus cloud deck when I became excited by the image of regularly spaced bands of echoes on the scope that I attributed to the clouds. Such tenuous clouds had not previously been detected. I called the navigator back to my scope to show him this wonderful discovery. He looked at me in disgust, then went back and lowered the radar antenna below the aircraft where it could view the outside world. Those beautiful bands of echoes were simply internal reflections from within the airplane. A few years later, this experience proved to be helpful when a conference speaker displayed similar bands recorded on an APQ-13 radar on board a B-29 aircraft off the coast of Japan. He attributed these to ocean waves, but the change in band orientation with the turning of the aircraft suggested that they were interference patterns resulting from the direct echoes and those received via reflection from the airplane wings.

A few weeks later, Col. Philpott suggested a flight to California to

7

test some advanced windshield wipers for the B-17. We had a "convenient" equipment failure that forced a layover, which we spent in Hollywood. Philpott had piloted a United Service Organization troupe of actors and actresses around the China–Burma theater during the war, so he had a number of good contacts in the movie community. It was a great thrill for this kid from Brooklyn to meet Clark Gable, Keenan Wynn, and other film stars of that day.

AWFD was moved to Clinton County AFB (still Army until 1948) at Wilmington, Ohio, by the end of 1945. There we instrumented a couple of B-25 aircraft with APQ-13 radars and simple meteorological instruments and undertook some elementary research. Among the novel instruments we flight-tested was the dewpoint hygrometer invented by Verner Suomi, then still associated with the University of Chicago. Vern was later to invent the scanning camera for geosynchronous satellites, and was a bottomless pit of innovative ideas. I was discharged in October 1946 and stayed on as a civilian. Following the lead of Stewart Marshall in Canada and Alan Bemis and Polly Austin at MIT, our initial work was aimed at quantitative radar rainfall measurements. This was soon to lead to my first significant development.

Another lucky break was the arrival of Major Joseph (Joe) O. Fletcher to assume the leadership of meteorological research in the early fall of 1945. Joe was the prime mover for the establishment of radar meteorology in 1943 (4) when he recognized the importance of radar for wind finding and weather observation. His key role at AWFD was to act as the liaison between the Air Corps, the Weather Bureau, and the academic community in the design and execution of the Thunderstorm Project (TP). The project's primary goal was aviation safety (5;2) This matched the key objective of AWFD, so it was a natural collaboration. As Joe's assistant I was able to attend the first major organizational meetings for the TP at the University of Chicago. All the major figures of the time participated—including Carl Gustav Rossby, Jacob Bjerknes, Francis Reichelderfer, Horace Byers, Henry Houghton, Ross Gunn, E. J. Workman, Alan Bemis, Harry Wexler, Henry Harrison (then of United Airlines), and others from the aviation community. I was awestruck by all the great names that I had previously seen only as authors of papers and books. In a later meeting, I presented a simple-minded description of the operation of radar beacon transponders to physics Prof. Michael Ference only to be informed afterwards by Prof. Byers that Ference knew a great deal more about transponders than I did. It was kind of him to listen to me so patiently.

My first task in connection with the TP was to visit the Army Air Corps Radar Test Center at Orlando, Florida to help select the radars to be used. To my utter surprise and pleasure, a hurricane occurred and moved directly north up the Florida peninsula on September 15–16, 1945. I was fascinated by the eerie calm and beauty of the sunlit clouds when the eye of the storm passed directly over us. Never having experienced such winds and not knowing whether the radars would withstand them, all but one of the radar staff abandoned me while I ran from the CPS-1 (microwave early warning—MEW) to the SCR-615 radar to take pictures of the scopes with a jury-rigged Speed Graphic camera for some 36 sleepless hours. I was entranced by the beautiful spiral bands and the ease with which I could plot the eye of the storm. (It wasn't until later that year that I saw the report of Maynard showing a few radar photos of a typhoon.) Unfortunately, my eagerness to analyze and publish the observations was frustrated when the photo record was received by the headquarters of the Air Weather Service and Major Harry Wexler published the paper (6) without knowledge of, or attribution to, the photographer. Of course Harry did a splendid job, although he and others failed to recognize the stratiform nature of the outer bands until some years later. Harry too became a good friend a few years later. Surely, the most important product of this experience was the seductive taste of the sense of discovery. Perhaps it was not yet fully within my consciousness, but I also began to recognize that one had to be opportunistic and flexible to exploit events when they occurred.

Another memorable event at the All Weather Flying Division was the invention of isoecho contour mapping. We had been working on the radar measurement of rainfall by using the stepped-gain method in which the radar sensitivity was reduced sequentially on successive antenna scans so that the storm boundaries at each step represented a contour of equal echo intensity (i.e., isoecho). But the process took too long, and I was wracking my brain as to how this could be speeded up. As often happens, the idea came to me when the brain synapses were relaxed at the bar in the Officer's Club at Clinton County AFB with my supervisor, pilot, and friend, Lt. Seymour Shwiller. I suddenly realized that I could *subtract* the signal above a specified threshold from that at a lower threshold so that the area in which the echo exceeded the higher threshold within the core of the storm would appear black. We implemented the scheme the next day and it worked like a charm. The advantages of the approach for the display of the internal structure of a storm on a single antenna scan and quantitative precipitation measurements were exciting. And it was immediately apparent that such a display in the cockpit would be of immense value to pilots in avoiding hazardous storms and turbulence. My excitement was multifaceted. I was thrilled by the benefits to science and aviation, but the thoughts of fame and fortune were equally prominent. And I was also mindful of my parents' pride in yet another achievement.

I prepared a patent disclosure and submitted it in October 1947. The patent (7) was granted in 1953 and a broader reissue patent (8) was granted in 1955. I particularly emphasized the application to airborne radar for the detection and avoidance of turbulence and severe storms. A few years later, the Stormy Weather Group at McGill University developed gray-shade mapping at multiple levels.

The technique received publicity in *Aviation Week*. It was subsequently incorporated into an experimental RCA radar, was flight-tested by United Airlines, and became operational in 1954. At that time, I brought suit for infringement. RCA settled the suit in 1958 and Collins Radio followed two years later. The first word from my patent attorney of RCA's willingness to settle reached me in the middle of the night at a U.S.-Scandinavian meteorological conference in Bergen, Norway in late June 1958. He asked if I would like royalties or a lump-sum settlement. Here I was suddenly to become rich, and all I could think about was the cost of this telephone call! Well, after a bit of discussion I told him "royalties." Of course, there was little chance of sleep thereafter, so I waited until about 5:30 A.M. to wake Lou Battan in the same hotel. I needed a confidant to talk to. This was the time of the white nights, so the sky was bright and early birds were already up and about. We walked and talked, but Lou wisely resisted giving me advice concerning the royalty–lump sum option. Nevertheless, I changed my mind and called my attorney back to tell him to take the lump sum on the basis of the "bird in the hand" theory. This turned out to our advantage two years later when we lost our suit against Bendix Radio because of a public use issue. The reason for this was that, in my enthusiasm, I had told my friend Frank White at American Airlines about the invention, and he implemented it on board an American Airlines experimental aircraft under a Navy contract. At the end of the contract, he invited reporters to fly in the aircraft. Although there were no storms to be viewed by the radar and the operation of the contour mapper was not demonstrated, the presence of the reporters was declared a public use. The flight took place in August 1949, just 13 months prior to the filing date of my patent (delayed because of the large backlog of wartime inventions). Had the flight occurred just a month later the question would have been moot. We appealed to the Federal Court of Appeals in Boston on the grounds that the flight was experimental, but the appeal was rejected by a vote of 2 to 1. Later the Supreme Court refused to hear the case.

Of course, we know that airborne weather radar quickly became a standard tool for all commercial and most private aircraft. And color-coded weather radar maps are everyday features of all weather broadcasts. It was most gratifying to receive a letter a half century later from a Uruguayan pilot expressing appreciation for the value of contour mapping to flight safety.

In passing, it is interesting to note that Vannevar Bush, the WWII tsar of the National Defense Research Board (NDRB), strongly recommended that investigators working under government sponsorship retain commercial patent rights. This was part of his proposed program for what was to become the National Science Foundation (NSF) in 1950, but was rejected by President Truman in an executive order in January 1950. Luckily, my invention predated that order.

To return to the Thunderstorm Project, I tried to convince Prof. Byers to modify the radars to provide for storm contour mapping when the project moved from Florida to Ohio in 1947. But he was reluctant to try this unproven approach at the last minute. Thirty-nine years later, in his last letter to me before his death in October 1986, Lou Battan expressed great regret that none of the radar images in the Thunderstorm Project report (5) are quantitative. I might note that Byers later became one of my greatest supporters and was instrumental in nominating me for the Meisinger Award of the American Meteorological Society, my first award from a professional society.

An incidental byproduct of the TP was the analysis (never published openly) of the depth of penetration of thunderstorm echoes below the horizon by a radar height finder. This was found to be a mirror image of the storms seen by reflection (from the sandy surface) at small negative angles of incidence. Since the reflection coefficient of the earth surface decreases with increasing negative angles, the deeper echoes require stronger storm reflectivity and greater rain rates.

It is difficult to overestimate the personal benefits of my association with the Thunderstorm Project. The project was the first major meteorological field program and became a prototype for many to follow. But the methodology of data recording both in the aircraft, by photographing the instruments, and on the ground, using old-fashioned ink and pen recorders, was primitive, and made data analysis extremely tedious. My appreciation of modern digital recording and computer analysis is heightened by recollections of the outmoded methods that we had to use in the early days. The entire TP program was an unprecedented learning experience. Horace Byers, Roscoe Braham, Harry Moses, Patrick Harney, and others became my friends and colleagues. I maintained close contact with Lou Battan throughout the analysis period and his subsequent Ph.D. dissertation, and indeed throughout his life. And I learned much more in later years as editor of AMS Meteorological Monograph No. 6 on severe storms, when Ted Fujita and Chester Newton published their classic papers based upon data from the project.

The combination of the first hurricane observations, the development of radar contour mapping and its impact upon aviation and rainfall measurements, and the association with the Thunderstorm Project sufficed to addict me to science. All three occurred while I was between the ages of 21 and 23. It was heady stuff for a young man. Once bitten by the bug I was hooked and sought ever more of that incomparable taste of discovery.

CHAPTER 4

Air Force Cambridge Research Laboratories, Geophysics Research Directorate—1948–1966

Section 4.1. Background

In October 1948, a few weeks after my marriage, I joined the Geophysics Research Directorate (GRD) of the Air Force Cambridge Research Laboratories (AFCRL) as chief of the Weather Radar Branch. A comprehensive account of weather radar at GRD has been presented by Metcalf and Glover (9), so I shall not dwell on technical details. Suffice it to say that we covered many of the facets of the field during my 18-year tenure there. Members of the Weather Branch during that period included Vernon Plank, Wilbur Paulsen, Harold Banks, Ralph Donaldson, Edwin Kessler, Keith Browning, Albert Chmela, Ken Glover, Ken Hardy, Pio Petrocchi, Roger Lhermitte, Roland Boucher, and a number of key visitors (Plate 3). Although not a member of our staff, Raymond Wexler (Harry Wexler's brother) spent most of his time at our laboratory (Plate 4).

During the early years, prior to the establishment of the National Science Foundation, GRD took the lead in funding a major portion of academic research in the atmospheric sciences, so we built strong relations with university investigators. In our case we developed excellent connections with Harvard Blue Hill Observatory, Tufts, Northeastern, Illinois State Water Survey, Texas A&M, and McGill University. Our relations with McGill were particularly strong; they are described later. Plank, Kessler, and I were also part-time graduate students at MIT, so we had regular exchanges with the MIT Weather Radar Project (10). MIT Lincoln Laboratory was also a neighbor, and provided us access to some remarkable advanced radar facilities. In later years, we also contracted with the Raytheon Company in Wayland for cutting-edge work on Doppler radar, and benefited from experts in signal processing such as Herbert Groginsky.

It is difficult to convey a feeling for the research atmosphere that prevailed a half century ago. All of meteorology was just emerging from the dark ages and radar was still an untapped tool. We were all young pioneers with unbounded enthusiasm and a wide universe of problems to explore. Management was loose, flexible, and supportive. We were not restrained by tightly specified research plans and could go in hot pursuit of targets of opportunity. As long as novel results were attained, financial support was forthcoming. This was especially true for the Weather Radar Branch, which managed to break new ground repeatedly.

The detailed account of our work provided in (9) summarizes our primary achievements, while the parallel histories in (1, 2, 3) put them into a broader context. Thus, I shall focus mainly on those scientific detective stories that were particularly intriguing and contributed most to the further development of the field.

But let me first recount our mutually stimulating relationship with the Stormy Weather Group at McGill University under the direction of Stewart Marshall. They had already made significant advances in quantitative rainfall measurements in the period 1945–1949, and earlier at the Canadian Army Operational Research Organization. Our interests at GRD were closely matched to those at McGill and I had quickly developed a close rapport with Stewart. Subsequent to a visit with him and his family (then living at Dorval Airport in close proximity to their radar), we arranged a research contract with the university for the partial support of their work starting in 1949. This support and close working relationship continued without break until 1964. In addition, I arranged for the provision of one of the first CPS-9 radars (I believe it was Serial No. 2) to McGill in 1953 and for the transfer of a brand-new 10 cm wavelength FPS-18 Doppler radar to them in 1964. The latter system remains the keystone of the McGill Radar Observatory in 2001. In the early years, I visited them three or four times per year, and spent most of the summer of 1955 as their guest. Similarly, Stewart, Walter Hitschfeld, and Ken Gunn each spent a month with us in 1952.

It is hardly necessary to recite the achievements of the Stormy Weather Group that were described so well by Dick Douglas (11). In the 1950s, it was difficult to distinguish our work from theirs, for we generally pursued similar lines of attack, usually following, sometimes leading. This included work on radar rainfall measurements, the nature of precipitation trails, the sorting of raindrops by wind shear, and a range of technological developments. I shall only describe a few studies that were strongly influenced by our interactions.

In 1951–52, Shepard Bartnoff, Roland Boucher, and I undertook two short studies of the relation of the radar reflectivity of clouds to the features of the drop-size distribution (DSD) and water content. We found that the reflectivity factor (Z) could be expressed as a function of some effective drop size such as the median volume diameter, the cloud water content, and a factor dependent upon the form of the DSD. (See the tutorial on basic radar meteorology in Appendix C.) This differed from the prior findings of Marshall and Palmer (M-P, 12) that established the radar–rain rate relation corresponding to an exponential DSD. But there were literally hundreds of such relations reported in the literature, and many of these differed greatly from that of M-P (13, 14).

Our early efforts led to the development of the first rain parameter diagram—RAINPAD (15). This showed that the radar–rain relation depended mainly upon the relation between the effective raindrop size and rain rate (R), which often differed greatly from that of M-P for stratiform rain in Montreal. It also explained the effects of evaporation and drop growth on these relations. Two decades later, the RAINPAD was generalized further (16) and showed how one could deduce microwave attenuation from any pair of parameters. Seliga and Bringi (17) introduced polarization differential reflectivity (Zdr)—supposedly a unique function of effective raindrop size. It was thus another parameter that could be combined with any of the others to estimate rain rate. Of course, this was hardly the end of the story, for Ulbrich (18) proceeded to show that the gamma function was often a better fit to the drop-size spectrum, and that truncation of the DSD further complicated the measurement of all the parameters. Recently, we have exploited this basic work to show how the $Z\text{-}R$ relations of tropical oceanic rain behave nearly systematically. That is, they can be classified into three types—convective, transition, and stratiform—based in large part on the behavior of the relation between drop size and rain rate (19).

But it would be unrealistic to think that radar rainfall measurements depended solely on the form of the drop-size spectrum. Zawadzki (20) pointed out the various errors in radar measurements such as those due to partial beam filling and sharp reflectivity gradients that often influenced the accuracy of remote rainfall estimates.

Although my interest in the nature of rain and its radar properties has spanned more than half a century, the literature is replete with studies of this subject. Our own work merely comprises a small fraction of the vast archive of information on precipitation and its particle size distribution. Research on this subject has burgeoned with the 1997 launch of the Tropical Rainfall Measuring Mission (TRMM) and the need for improved algorithms to measure rainfall from space and to validate the measurements at the surface.

Section 4.2. Angels and Clear-Air Turbulence

Perhaps the most intriguing enigma of the first quarter century of radar meteorology was the story of angels—those mysterious echoes of unknown origin. The Weather Radar Branch at AFCRL debated their origin ad nauseam and devoted a great deal of effort to their study. We saw angels of all sorts with our vertically pointing 1.25 cm wavelength radar. The subject was not new, for radar investigators had reported unexplained echoes from the earliest days of radar. Vernon Plank (21) undertook a masterful study of the subject that includes an excellent history. Among other things, he credited Watson-Watt et al. (22) with the first observation of layer-like echoes at 50 m wavelength from heights below the ionosphere. (Watson-Watt, Wilkins, and Bowen were later credited with the invention of radar in the United Kingdom.) When I repeated this report in my own review paper (23), I received a personal note from Sir Edward Appelton (dated 15 July 1959) suggesting that the echoes reported by Watson-Watt must have been from ground clutter. Appelton and Piddington (24) had conducted experiments at 34 m wavelength and concluded that the echoes of Watson-Watt et al. (22) could not have been from ionized layers or patches; rather, they must be due to atmospheric scattering from regions of very low reflection coefficient and from ground clutter. According to Plank, it was Friend (25) who presented the first persuasive evidence of stratified echoes from within the troposphere.

Following WWII there was a debate of some 25 years duration concerning the origin of "dot angels," angels of short duration and very small size. An early study by Crawford at Bell Telephone Laboratories concluded that they were due to insects (26). However, because their diurnal and seasonal behavior mimicked that of convective thermals, a number of us continued to suggest that they were due to reflections from the thermals themselves. I am afraid that I contributed to the confusion and the sometimes heated, often amusing arguments on the source of these echoes. In 1959 (23), I developed a highly simplified theory for the radar cross section of spherical thermal "blobs" that purported to explain their echo strengths; however, this proved erroneous in the light of later developments. My 1964 paper "Angels in focus" (27) (of which I was quite proud at the time) attributed dot echoes to the caps of thermals where the sharp discontinuity in moisture and refractive index would be aided by the focusing effect of the shape of the cap. The main benefit of that work (which was ridiculed by Stewart Marshall at the 11th Radar Meteorology Conference) was that it permitted us to establish a hypothesis that could be tested when we obtained powerful multiwavelength observations as we did a few years later at Wallops Island. This is an archetypal example of the uncertain and erroneous theories that result from the lack of adequate observations.

Oddly enough, when we finally obtained adequate ultrasensitive radars, we found that both insects and the boundaries of clouds and thermals were responsible for the echoes. Observations and discussion are presented in the comprehensive review by Gossard (28), where one sees hordes of insects carried by the convective and wave motions. Wilson (29) attributes the vast majority of the echoes at wavelengths of 10 cm or less to insects. Evidently, the reason that it is often difficult to distinguish between insects and atmospheric echoes is that winds and convection modulate the patterns of swarms of insects so that they resemble atmospheric phenomena. Examples of echo patterns that cannot be attributed unambiguously to either atmospheric phenomena or insects may be found in Figure 2.3 of Gossard's review (28). So, the argument goes on.

But we are getting ahead of ourselves. During WWII, there were many reports of microwave propagation beyond the horizon. Many of these could be attributed to anomalous propagation associated with the refraction of the rays by the lapse of refractive index with height. However, there were others that implied forward scattering by elevated layers in the atmosphere. In fact, in the 1950s the military had developed a major microwave communications system based upon scatter propagation. So it was not surprising that there was widespread interest among the radio communications community in the mechanisms responsible for such signals. In 1955, a voluminous special issue of the *Proceedings of the Institute of Electrical Engineers* was devoted to this problem.

Meanwhile, radar meteorologists were pursuing their interests in angel echoes. Let us return to the early summer of 1953 when we conducted an experiment at the MIT Round Hill Field Station in South Dartmouth, Massachusetts to study the detection of fog by 1.25 cm radar. While calibrating the radar later in the morning after the fog dissipated, we were surprised by mysterious angel echoes moving in toward the shore where the wind suddenly switched toward the sea, the temperature dropped, and the refractive index increased. And when using the vertically pointing beam we detected quasi-coherent echoes from the sea-breeze inversion (30). These observations elicited great excitement when presented at the Fourth Radar Meteorology Conference in 1953. However, the experts remained skeptical because the subcentimetric scales of refractive index fluctuations or gradients that were required to explain the echoes were supposed to have been eliminated by diffusion (Appendix C). But no one could offer another explanation. In any case, the sea-breeze observations were a classical case of serendipity and were among those intriguing mysteries that drove us on. A half century later, I am inclined to believe that the echoes could not have come from the sea breeze itself because it is very unlikely that half-wavelength (0.6 cm) perturbations in refractive index could have existed in the presence of dissipation. Instead, it is possible that refraction of the radar beam toward the sea surface detected the capillary waves on the sea generated by the sea-breeze winds. See the discussion of synthetic aperture radar in the tutorial in Appendix C.

A few years later Harper et al. (31) used their 10 cm height finder in England to observe inverted U-shaped echoes in vertical cross section corresponding to the boundaries of cumulus clouds; these were called *mantle echoes*. That year we also detected cumulus mantle echoes with the powerful 10 cm FPS-6 radar height finder operated by MIT Lincoln Laboratories at South Truro on Cape Cod. But this was only possible after I expanded the scope to display the first 20 miles (23). Evidently these weak echoes had not been seen earlier because the radars were generally operated at long ranges for the detection of aircraft and the weaker echoes were compressed near the origin of the display where they were obscured by ground clutter and ignored as artifacts. The nature of the mantle echoes was strongly supported by aircraft penetrations of cumulus clouds that recorded the sharp discontinuities in refractive index, and the implicit small-scale fluctuations required to account for the echoes (32). It is noteworthy that it took another 40 years for Knight and Miller (33) to confirm that mantle echoes were due to Bragg scatter from refractive index turbulence rather than to scatter from rain or cloud drops (Appendix C).

We also observed layered echoes associated with sharp fluctuations in refractive index measured simultaneously by an aircraft traversing the inversion at the same height as the echoes (23). Similar correlations between the heights of stratified echo layers and aircraft-measured discontinuities in refractive index were reported by Lane and Meadows (34) in England. Thus, there was increasing evidence that sensitive radars at appropriate wavelengths could detect clear-air structures.

But it was not until Tatarski (35) developed the comprehensive theory of wave propagation through a turbulent atmosphere that the subject could be addressed in a realistic and quantitative manner. This and the work of Smith and Rogers (36) was a point of departure for the assessment of the feasibility of the detection of clear-air turbulence (CAT) by ultrasensitive radar (37). The latter study was well under way in the 1963–64 period at a time when the Air Force was deeply concerned with the dangers of CAT. Coincidentally, it was also the time when MIT Lincoln Laboratory had completed its radar studies of the echoes from ballistic reentry vehicles with three ultrapowerful radars (at wavelengths of 3, 10.7, and 70 cm) at Wallops Island, Virginia. We recognized the unique opportunity to use these radars and proposed that the Air Force and NASA assume joint responsibility for their support and operation. The result was a multimillion-dollar grant to conduct studies of CAT and related clear-air phenomena. We contracted with a team of researchers headed by Isadore Katz of the Applied Physics Laboratory of Johns Hopkins University to join us in this work. It was a coincidence that the first radar detection of CAT at the tropopause (38) occurred just a few months prior to the 1966 publication of the feasibility study.

The detection of the tropopause was only one of the great variety of breathtaking and barely believable discoveries that flowed from the work at Wallops Island (39, 40, 41). These included lines of donut-shaped convective cells, corresponding to horizontal cuts through mantle echoes, Kelvin-Helmholtz waves (like breaking waves on the ocean) which were later found to be the source of clear-air turbulence, boundary layer inversions, and the characteristics of individual insects and birds (42).

The unprecedented observations at Wallops Island opened the door to a broad range of research because it permitted us to view atmospheric structures in three dimensions, which had either never been seen before or could barely have been anticipated from radiosonde or aircraft measurements. The history of subsequent developments is described by Hardy and Gage (43). They also recount how the work at Wallops Island stimulated the steps taken to use longer wavelength radar, previously devoted to ionospheric research, for studies of the troposphere in the early 1970s. This initiated the era of mesospheric-stratospheric-tropospheric (MST) radar that has provided the basis for much of our understanding of the dynamic links between the lower and upper atmosphere.

And then came the VHF and UHF wind profiling radar that provides virtually continuous measurements of the winds directly overhead. Who could have guessed at this phenomenal evolution when we first began to explore the origins of angel echoes a half century ago?

The 1957 conference at which Plank and I presented the work published in the *Journal of Atmospheric and Terrestrial Physics* (23, 32) had been organized by Stewart Marshall and Bill Gordon and was attended by a number of the experts in radio propagation and meteorology at the time. The resulting cross-fertilization was so productive that Marshall wrote:

> We submit, therefore, that a Joint Commission on Radiometeorology (of URSI and UGGI*) should be retained, with membership from these two Unions, and charged with the responsibility to maintain constructive intercommunications among radiometeorologists, whether their basic interest lies with radio or meteorology, or purely and simply with the composite appellation. [*URSI is the International Radio Science Union; UGGI is the International Union of Geodesy and Geophysics.]

It was this proposal that gave rise to the formation of the Inter-Union Committee on Radio Meteorology (IUCRM). IUCRM held a long series of conferences on a variety of subjects of mutual interest. The meetings held in Moscow (1965), Stockholm (1969), and La Jolla (1972) were pivotal in the explosive growth of our understanding of clear-air scatter and the meteorological processes that I have described above

(2, 43). These gatherings, attended mainly by invitation to those we thought would stimulate the discussions, were models of truly creative meetings.

Section 4.3. Back to the Early Days

We now return to 1957. As chairman of the AMS Committee for Radar Meteorology, I led the organization of the Sixth Weather Radar Conference at MIT in 1957. During the banquet, I introduced Dr. Robert Fletcher, president of the AMS, to say a few words. He had been sitting next to my wife Lucille at the head table and hinting to her of the travels that we would be able to afford after that night. When he spoke, he presented me with the Leroy Meisinger Award and a check for $100 (the same amount that still accompanies the award today). I was flabbergasted. Having had a couple of martinis I was well lubricated and expressed my appreciation in words such as these:

> I want you all to know that this award belongs to you as much as it does to me, for it would not have been possible without all the ideas which I have stolen from you. But my conscience is perfectly clear about this for I know that you have stolen at least as many from me.

I could say that then because the small contingent of attendees (about 70) were all my friends and we were used to poking fun at one another.

The Seventh Weather Radar Conference was held at the Deauville Hotel in Miami Beach in November 1958. Plates 5, 6, and 7 show the mostly young radar meteorologists at the outdoor banquet. These photos are heartwarming for they remain among the few mementos of the pioneers who have left us.

Another significant collaboration with the Stormy Weather Group was the work on polarimetry that I did with Milton Kerker and Walter Hitschfeld (44). Milton Kerker, a colloid chemist and an expert on particulate scattering, had spent some time at McGill in 1949–1950 and had done work concerning scattering by nonspherical particles based upon the 1912 work of Gans. (The latter paper dealt with the form of ultramicroscopic gold particles.) With this lead, I did a theoretical study of the properties of the backscatter from various hydrometeors for my master's thesis at MIT. Later I joined forces with Kerker and Hitschfeld in the formal publication. I quote one paragraph that speaks to the essence of the results:

If the particles have preferred orientations, the back-scattered energy depends not only on the particle shape but also on the polarization of the radiation and on its angle of incidence. This dependence is more pronounced for water (or water-coated ice) particles than for ice, while snow, on account of its low refractive index, scatters almost like spheres of equal volume, regardless of particle shape or orientation.

This extract only hints at a great deal of what we have learned since. I refer you to the comprehensive review by Seliga et al. (45). Some work had been done on polarimetry in England by Robinson and Browne, by Labrum in Australia, by McCormack and Hendry in Canada, by Barge and Humphrey at McGill University, and by Newell and his colleagues at MIT. However, it was not until Pruppacher and Pitter (46) found that the ellipticity of raindrops increased with their diameter and that their large dimensions were horizontally oriented that one could expect differential echoes from horizontally and vertically polarized radar waves. This was implemented by Seliga and Bringi (47) and stimulated the entire field thereafter. The advances made since 1976 are revolutionary.

Because Metcalf and Glover (9) present a comprehensive review of the research program at GRD, I shall only describe the background of some of our other endeavors. As noted earlier, we did not rely solely on our own radar facilities; rather, we exploited those that were better suited and positioned to observe special weather events, such as the 23 cm wavelength FPS-20 search radar and 10 cm FPS-6 height finder operated by MIT Lincoln Laboratory at South Truro, Massachusetts.

We used those radars for intensive observations of Hurricane Edna in 1954 (51) and of Hurricane Esther in 1961 (52). Those were unique opportunities. Few hurricanes have reached New England in the decades since. The first of these reports was wide-ranging and covered virtually all features discernible by radar. The second focused on the stratiform nature of the outer bands and pointed out that the precipitation originated in the convective cells at the upwind end of the bands. Subsequently the latter structure became the model of tropical rainbands not necessarily associated with hurricanes (53).

During the Hurricane Edna episode, I called in the positions of its eye to a colleague at Lincoln Lab. I was also listening to the radio reports of its position from the National Weather Service, presumably based upon the Air Force reconnaissance observations, and was excited by the accuracy of my reports. When I returned to Boston two days later I discovered that the Weather Service was using my positions, which were relayed to them from Lincoln Lab. That's one way to achieve great accuracy.

Lucille did not appreciate my forays to Cape Cod to monitor hurricanes. On each such occasion I would dash home with a large pack of dry ice, candles, and flashlight batteries in the event of a power outage, leaving her to care for the children.

During one of our periodic visits to South Truro, another bit of serendipity led to the detection of lightning echoes extending vertically from the ice phase echo tops of thunderstorms to higher altitudes in the clear air where precipitation echo was absent (54). Herb Ligda of the MIT Weather Radar Project had previously seen lightning echoes, but none had been observed on a height-finder. The recently discovered blue jets over an active hailstorm (55) suggest a possible connection between the blue jets and radar lightning observations. We were also remarkably lucky to have one of our cooperative observers positioned under the core of that storm, where he reported the time history of the number of lightning strikes to ground as the storm passed overhead.

Perhaps one of our most pivotal contributions resulted from the invitation from Lincoln Laboratory to use the 5.6 cm Porcupine Doppler radar installed on the roof of their Lexington, Massachusetts building. This occurred during the visit of Dr. Roger Lhermitte of France to our laboratory in 1957. Roger attached an audio speaker to the analogue Doppler output and we rotated the beam in azimuth during a rainstorm. To our astonishment and exquisite pleasure on 2 Dec 1957, we heard and tape-recorded the Doppler shift as it varied in pitch from near zero frequency when the beam was pointed crosswind to high frequencies when pointing either up- or downwind. We heard the wind! Within a few days, we had borrowed a Rayspan audio analyzer from the Raytheon Company and could display the Doppler frequencies or velocities as a function of azimuth. Thus was born the velocity-azimuth display (VAD) (56, 57). The VAD allowed the measurement of the vertical profile of winds in precipitation. Plate 8 shows Roger Lhermitte with me and AFCRL Commander General Ben Holzman when we were recognized for the VAD patent. The complete VAD theory was subsequently developed in an elegant paper by Browning and Wexler (58).

Roger Lhermitte is one of the brightest stars in the weather radar firmament. He can do it all, from the basic engineering to the sophis-

ticated meteorological analysis. Among his recent achievements is the
development of a 94-GHz Doppler radar. This short wavelength makes
it possible to detect the maxima and minima in the Doppler spectrum
of precipitation according to the Mie theory, and thus allows the
recovery of both the vertical air motion and the drop-size distribution
in rain (Appendix C). Pavlos Kollias, a postdoctoral research associate
at the University of Miami, wrote a superb dissertation that revealed
heretofore-unknown features of the evolution of the precipitation
in both convective and stratiform rain. At age 80, Lhermitte is
nearing the completion of what appears to be a major book on radar
meteorology.

    This account does not mention the pioneering activities in Doppler
radar that were being done by Brantley, by Rogers and his colleagues
at Cornell Aeronautical Laboratories, Lhermitte in France, Battan,
and Theis at the University of Arizona, and by Boyenval and Caton in
England. I refer the reader to Rogers (59) for an exciting account of
that work.

    The Porcupine radar was transferred to the Weather Radar
Branch in 1961 and became the primary tool for a broad range of basic
studies. Among these was the development of the plan shear indicator
(PSI) by Armstrong and Donaldson (60), with which Donaldson determined the pattern indicative of a mesocyclone prior to and during a
tornado. This was one of several important developments that laid the
foundation for the development of the next generation weather radar
(NEXRAD). I refer you to a series of papers on the use of Doppler radar
signatures for severe storm detection and warning by Ralph Donaldson
in the list of references in *Radar in Meteorology* (3).

    By the late 1950s, it was obvious that Doppler radar was going to
be the basis for any future weather radar. In 1957, I visited Dr. Francis
Reichelderfer, the director of the U.S. Weather Bureau, to persuade
the Bureau to incorporate Doppler techniques in the WSR-57 radar,
which was then in the design stage. In a talk at a dinner meeting at the
Raytheon Company in 1963, I also proposed that Doppler methods
were the only solution to the tornado detection problem. But the
increased costs were thought to be prohibitive, and so we were to live
with the incoherent WSR-57 for some 30 years before NEXRAD came
along. In fairness, it must be noted that digital Doppler measurement
and visualization techniques such as the pulse pair processor and color
displays were not yet available, so that operational Doppler radar was

not yet ready for development. It was the revolution in digital signal processing techniques in the 1970s and 1980s that completed the mix of technological ingredients to make NEXRAD a reality.

One of the primary activities of the Weather Radar Branch in the 1960s was the work by Donaldson and associates and by Browning, who joined us in 1962 after completing his Ph.D. at Imperial College. As we discuss below, he and Ludlam provided fundamental new insights on the structure of severe thunderstorms (61). Working in collaboration with the National Severe Storms Laboratory (NSSL) in Oklahoma, Donaldson and Browning conducted a number of landmark studies of the nature and evolution of tornadic storms (see 9 for references). By 1966, I had left to join the University of Chicago and Browning returned to England. But the members of the Weather Radar Branch, in conjunction with their associates at NSSL, persisted in their pursuit of an operational Doppler radar. With the growing recognition of the operational value of Doppler radar by the National Weather Service, the military weather services, and the FAA, these efforts gave rise to the Joint Doppler Operational Project (JDOP) conducted at NSSL during 1977–1979 (62, 63), and JDOP ultimately led to the establishment of the NEXRAD program.

Section 4.4. Personal Reflections on Stewart Marshall and Walter Hitschfeld

A necrology of Stewart Marshall appears in the AMS *Bulletin* (48). We all learned a great deal from Stewart. His scientific intuition and imagination were impressive—and the productivity of the group was second to none. With collaboration by people like Walter Hitschfeld, Ken Gunn, Tom East, and Dick Douglas, his qualitative ideas were turned into remarkable studies such as the classical paper on the interpretation of the fluctuating echo from randomly distributed scatterers (49).

One of my early experiences with Stewart was at a meeting in Washington, D.C. in 1950 when he, Walter (and, I believe, Ken Gunn), and I were attending a meeting on precipitation. He took me under his wing like his students and rehearsed our presentations until the wee hours of the morning. His emphasis was on focusing our talks on just one or two points. He himself was a superb speaker who got his message across smartly. But Stewart could also be a punishing chair-

man and critic, wielding his tongue like a sword to cut speakers down
mercilessly. It was not pleasant to be the target of such an attack, as
I can testify from personal experience. On the other hand, he was the
epitome of the proper British gentlemen when I was a dinner guest at
his home with his charming wife Beth and daughters Claire and
Heather.

The obituary for Walter Hitschfeld also appears in the AMS *Bulletin* (50). He died in 1986 at the young age of 64. He was a terrific
scientist and a sweet human being. He was the keel of the Stormy
Weather Group and often the rudder that stabilized the group under
the sometimes erratic leadership of Stewart Marshall. His versatility
ranged from cloud physics to infrared radiation, the physics of hail,
and weather modification, and, of course, through the entire domain of
radar meteorology. While Stewart was often fiery, Walter was always
gentle. His judgment was impeccable and he exerted it effectively as
dean at McGill, in the AMS, at NCAR, and in international bodies. He
was universally liked and respected for his wisdom, wit, and good
humor. I miss him still.

Section 4.5. My Sabbatical in England—1959–1960

I was awarded a senior post-doctoral fellowship by the National
Science Foundation in the early spring of 1959 to work with Frank
Ludlam at Imperial College, London. My family—including my wife,
Lucille, nine-year-old Joan and five-year-old Robert—and I arrived on
the Queen Elizabeth I on June 2. We drove a rental car—nervously—on
the "wrong" side of the narrow roads from Southampton to our
charming Trust house at Little Gaddesden. The hotel was close to
the palatial estate, Ashridge House, which was the residence of the
research group that Ludlam had assembled to study thunderstorms
and hail. Our research headquarters was the Meteorological Office
Radar Station at East Hill, Dunstable, which was under the direction of Bill Harper.

Most of the group are shown in Plate 7. Missing from this picture
are Bill Harper and his technicians Coop and Denham, and Pat
Harney and Pio Petrochi, who accompanied me from AFCRL for the
field experiments.

I joined the research team in the luxurious Ashridge House during
the second month of the project there after my family settled in our

residence at Sunningdale. Each evening we left our shoes at the door to be polished and each morning we had tea and biscuits delivered to our rooms. And Frank Ludlam whipped all of us at snooker after dinner each night.

In recognition of Ludlam's distinction in studies of clouds and storms, I had arranged a research contract from the Air Force Cambridge Research Laboratories to support him and his students. In addition, in anticipation of the experiments to be conducted, I had organized the transfer of a 4.7 cm wavelength MPS-4 height-finder radar to Imperial College.

Following the field program, Ludlam had a tower constructed for the use of the MPS-4 radar at his Cheapside laboratory near Ascot. However, just before it was put into operation, a representative of the Queen ordered that the tower and radar be removed because it detracted from the view from Her Majesty's box at the Royal Ascot Racetrack.

The summer of 1959 was one of the warmest and clearest on record. We were rather naïve to plan a storm research program in England in any year. But luck again favored us—1959 turned out to be the exception of the century. But what could we do while waiting for a storm?

Having remembered that Ryde (64) did not extend his calculations of the cross sections of ice spheres beyond a diameter of one wavelength, I suggested that this was a great opportunity to validate and extend our knowledge of hailstone cross sections experimentally (Appendix C). For this purpose, Macklin made artificial 3-inch diameter hailstones by freezing water in tennis balls. These were then suspended in a lady's hairnet from a tethered balloon. About 150 ft below the hailstone we suspended a 12-inch metal sphere as a reference standard target with known cross section. We could then tilt the beam alternately between the unknown hailstone and the standard target to measure the difference in cross section with accuracy comparable to that achievable in the best laboratories at that time. The only difficulty occurred when, unbeknownst to Ludlam and myself, Browning and Macklin amused themselves and confused us by substituting tomatoes for the hailstones. The measurements were made at three wavelengths simultaneously—3.2 cm with the TPS-10, 4.7 cm with the MPS-4, and 10 cm with the Ames type 13 height-finder radars.

We found that large ice spheres scattered about an order of magnitude better than water spheres of the same diameter. This was a surprise because we all nurtured the Raleigh scatter law that said that (small) ice particles scattered about one-fifth as well as equal water drops. Until then, the radar meteorological community had either forgotten or ignored the 1941 findings of Ryde (64) that indicated that ice became a better scatterer than water at a diameter of about 0.6 wavelengths. I immediately sent a letter to Lou Battan at the University of Arizona informing him of our results. By sheer coincidence Ben Herman, then a doctoral student working on the scattering of infrared radiation by cloud drops (under Lou Battan), had programmed the Mie equations for the then state-of-the-art IBM 720 computer. So, Lou suggested that Herman run the program for ice spheres at microwaves. I received the theoretical results within a couple of weeks and plotted them along with the experimental measurements and returned them to Lou. In a tape-recorded message to me on October 8, 1986 (just three weeks before his death), he said:

> . . . and I must say, it was one of the big thrills of my scientific experience when I got your letter in which you had plotted your actual measurements against the curve. I remember thinking to myself: isn't this astounding, theory and measurements are in agreement. (See my tribute to Lou Battan, 3).

Well, Lou's thrill could not have been any greater than that which Ludlam, Harper, Macklin, Browning, and I experienced 6000 miles away. It was just one of many examples of the fortuitous confluence of separately derived ideas that marks much of science. This work then became the motivation for an impressive list of studies conducted by Herman at Arizona and by my colleagues and myself at the AFCRL (14, 57).

But much more was to be done at Dunstable while waiting for storm activity. We also conducted experiments to check Harper's claim that ring angels (14) were due to starlings and not to some atmospheric phenomenon. To do this we sent Macklin and Browning out with walkie-talkies (at about 3 A.M.) to the tree roosts that Harper had identified as the sources of the ring angels. And lo and behold, just as the sun rose above the horizon, we saw the echoes (14, 55). As if on the signal of a wing commander, thousands of starlings took off simultaneously, spreading out in all directions, to have their insect breakfast in flight; they then formed nearly perfect circles, presumably in the innate knowledge that this was the way to fly in order to minimize the competition for their food. Observations such as these were to become the subject of the book *Radar Ornithology* by Eastwood (65).

Still another activity was the radar calibration of the 10 cm Ames type 13 and the 4.7 cm MPS-4, both in height-finder mode. The method was simplicity itself. We tracked a 24-inch diameter metal sphere as a standard target and calibrated the dial of the receiver gain control at which the echo from the sphere just reached the threshold of visibility on the scope. The threshold level was then a measure of the equivalent radar cross section of any target. The calibrations were performed on 8 July just in time for the Wokingham storm on 9 July (66).

The story of the Wokingham storm is legend. Here I shall use the succinct description of Probert-Jones (67).

On 9 July 1959 a severe storm developed over Brittany, crossed the channel near the Isle of Wight, and moved northeast across southeast England, passing just east of the radar site. That storm, the Woking-ham storm, was the last severe storm to cross southeast England for at least several years; fortunately for Ludlam and his team, its orientation, heading straight for the radar site, enabled features to be observed which otherwise would have been impossible to see. The radars were operating perfectly (all had been calibrated only the day before), and the radar operators each had been given a specific task which they carried out for several hours as the storm approached and moved away from them. The electricity supply failed for 45 minutes due to a lightning strike, but this occurred when the storm was almost overhead and radar observation would have been difficult in any case.

The analysis of the Wokingham storm was mainly carried out by Ludlam and his postgraduate student Browning (61). Ludlam had previously used RHI radar observations as well as visual measure-ments to obtain the rate of rise of echo or cloud tops to indicate vertical air velocities that could be related to hail size and thermodynamic structure. In 1959, one 3-cm RHI radar was specifically assigned to follow echo column tops to obtain their rate of rise. To his astonish-ment [and that of Browning and me] for a period of about an hour during the Wokingham storm it became impossible to discern any individual rising columns; throughout the period the echo mass moved with a uniform velocity and without any marked variation in its character. This led Ludlam [and Browning] to propound his [their] revolutionary concept of the steady-state model for the severe storm. Together with Ludlam and Browning's observation of particular fea-tures of the Wokingham storm (the wall, the forward overhand, and the echo-free vault), the steady state hypothesis enabled them to produce a three-dimensional kinematic model of the airflow within a severe storm.

Thus the analysis of the Wokingham storm was not only the culmination of years of acquired experience leading to a particularly comprehensive and well-organized observational program. It was also the beginning of an era in which radar observations would be allied to dynamical models to investigate classes of atmospheric circulations from severe storms through mesoscale phenomena to fronts and cyclones.

Note that I have emphasized Browning's role (in brackets) in formulating the steady-state model and the radar features such as the overhang and echo-free vault. These characteristics emerged only after many months of sometimes heated discussions during teatime at the Cheapside/Ascot laboratory and residence of the Ludlams. I have a vivid memory of 20-year-old Keith (Browning) insisting, in the face of friendly skepticism by Ludlam and me, that the updraft entered through the echo-free vault at the front of the storm and penetrated the intense echo to the back of the storm before turning around to exit in the anvil at the front.

During his subsequent (1962–1966) four-year stay at my laboratory (then located in Sudbury, Massachusetts) Keith Browning elaborated and solidified his concept of the structure and airflow in tornadic storms and introduced the term *supercell* to describe this class of severe storms. See the Browning references in *Radar in Meteorology* (3). In so doing, he set the course for storm studies for the rest of the century. He also conducted a wide range of other seminal research (9). Upon his return to the United Kingdom in 1966, Keith established a powerful program in radar meteorology at the Radio and Space Research Establishment (RSRE) at Great Malvern under the auspices of the Meteorological Office. That activity has since grown and given rise to a number of offshoots at several institutions so that the discipline is in a vigorous and healthy state there. During my tenure as director of the National Hail Research Experiment at NCAR in 1974–1975, Keith rejoined us for a year as chief scientist and again made a landmark contribution to the nature of the hail generation process (68).

Keith Browning is one of the all-time great meteorologists. He quickly learned all the tricks of the trade from his mentor, Frank Ludlam, extended them with his inimitable imagination and unmatched skill in depicting complex physical processes in exquisitely simple diagrams, and backed them up with rigorous mathematical analysis. While I like to think that I was his mentor in radar, it was not long before our roles were reversed. He was elected a Fellow of the Royal Society at the remarkably young age of 39.

My year with Ludlam will never be forgotten. He was a natural philosopher and a born observer with an uncanny ability to interpret what he saw in the clouds in terms of the physical processes within. He taught me to see the atmosphere in an entirely new way. And he encouraged us to make that intuitive leap that jumped observational gaps. I quote from the dedication that Sir John Mason (director general of the Meteorological Office) and I wrote to Ludlam's book *Clouds and Storms.*

Frank's penetrating insight into the workings of the atmosphere and his ability to piece together a physical picture of a complex physical mechanism from a few observational facts were matched by his great qualities as a lecturer and writer. In both his talks and his papers, he was able to translate intricate physical processes into elegantly simple descriptions which call up images in the mind of the reader. He was also able to convey his own appreciation of the beauty of clouds and weather through equally beautiful word pictures of unique clarity. And where words alone were inadequate, he resorted to illustrations of his own making which combined his skills as scientist and artist. In fact the bulk of the drawings in this volume were meticulously done by Frank himself.

Frank Ludlam died at his home near Ascot in England on June 3, 1977 at the age of 57 after an extended illness. I visited him often in the years before his death and was amazed at his continued sense of humor and his intense desire to debate the latest research findings. It was a special privilege for me (with the cooperation of Dean Charles Hosler) to arrange for the publication of his book *Clouds and Storms* by the Pennsylvania State University Press in 1980. While he never saw that book in final form, he was delighted to see early galley proofs just a few weeks before his death.

A footnote to this story concerns my interactions with John Saxton (director) and John Lane at the Radio Research Station at Slough, just a few miles away from the Ascot laboratory. There we had a number of stimulating discussions on the subject of clear-air scatter (i.e., angels), particularly the extremely sharp gradients in refractive index that were required to explain specular reflections such as those that we observed from the sea-breeze inversion (30). Whether or not this influenced their work, Lane subsequently developed a fast-response, three-dimensional refractometer capable of measuring differences over 0.1 and 1 meter vertically and 1 meter horizontally. At my invitation, in 1966, he brought this instrument to Wallops Island, where we suspended it from a helicopter and used it to determine the refractive index gradients and fluctuations in the clear-air echoes. It was in this way that Kropfli et al. (69) were able to confirm that the intensity of the echoes agreed with the Tatarski theory (28, 35).

It is impossible to track the subsequent evolution of the research activities in which I participated or the people and the institutions with which I interacted during that memorable year in England. But I cannot overemphasize the importance of periodic visits to other laboratories to revitalize one's thinking and initiative.

After leaving England, we went on a three-and-a-half-month, 10,000-mile driving tour of Europe with our children. It was an unforgettable trip, but out of context with this story. However, one of our stops was at the U.S. Air Force base at Torrejon, outside Madrid. We stopped there at the request of General Jack Taylor, my former commanding officer at the All Weather Flying Division, who was now in charge of air traffic control in the European theater. The radar traffic controllers there were having problems distinguishing aircraft from unknown radar angels. Somehow, General Taylor had learned that I knew something about the subject, and tracked me down in England. After a little study, I suggested that the radar operators make measurements of the strength of the echoes by successive reduction of the radar gain control until the echoes just disappeared from the scope. The resulting echo statistics were consistent with those from birds, i.e., roughly equal to the size of water spheres of the same dimensions. A quick fix was simply to modify the receiver sensitivity time control (i.e., sensitivity as a function of range) so that only the stronger echoes from aircraft would appear on the display. It was a simple solution but nevertheless a source of gratification that our academic research on angels had some practical application.

Section 4.6. The Soviet Union

One of the intriguing trips we made prior to my departure from AFCRL was to a meeting in Moscow in June 1965 on atmospheric turbulence and radio-wave propagation. This was a small conference that brought together experts on microwave propagation, turbulence, and fluid dynamics, and radar specialists interested in angel echoes. There were perhaps 25 foreigners and some 30 Soviet scientists. The meeting was a landmark that led to the subsequent conferences at Stockholm in 1969 and La Jolla in 1972 that I mentioned earlier. I was thrilled to meet Academician Alexander Obukhov, the head of the Institute of Atmospheric Physics and the successor to Kolmogorov, the father of turbulence theory; Prof. Akiva Yaglom, famous for his work on hydrostatistical fluid mechanics; and Valerian Tatarskii, a remarkable young theoretician in electromagnetic propagation, who also participated. I also met my Russian counterparts in radar meteorology, Andrei Gorelik and Albert Chernikov, among others. The mix of scientists led to the revolution in our understanding of scatter propagation beyond the horizon, radar angels, and waves and turbulence.

I was surprised and disappointed when we failed to hear anything from or about Tatarskii after the latter meeting until he showed up in the United States working at the NOAA Environmental Technology Laboratories in Boulder, Colorado in 1990. Nor was he to be found at the 1971 UGGI meeting in Moscow. He and Yaglom were punished for having signed letters protesting the imprisonment of some people accused of anti-Soviet activities and others who had been confined in mental institutions. The penalty paid by Yaglom and Tatarskii was a prohibition from all foreign travel, including nations of the Soviet bloc. Yaglom also lost his professorship at Moscow University. He informs me that the magnitude of punishment depended mainly upon the local communist party organization, and their penalties were relatively mild in the Institute of Atmospheric Physics.

It is noteworthy that the $-5/3$ power law for the turbulent energy spectra of velocity and scalar fields of temperature, humidity, and refractive index is due to Obukhov. This led Tatarskii, Obukhov's student, to the development of the theory of scattering of electromagnetic waves in turbulent media and the radar detection of clear-air echoes, discussed in Section 4.2 of this chapter.

There are a host of memorable anecdotes to recount about this visit to the Soviet Union at the height of the Cold War. Lucille and I arrived on a flight from Oslo along with our Norwegian friends Dag Gjessing and his very pregnant wife, Turil. We arrived about midnight and were greeted by a scene from a B movie. A grim Russian officer with his coat

draped over his shoulders boarded the plane and collected our pass-
ports, thus aggravating the apprehension that we had harbored since
leaving the States. We waited a nervous hour until the customs officers
searched selected luggage for banned literature and cleared us for the
long bus ride to Moscow in the company of our attractive interpreter/
baby-sitter, Galia. She was unexpectedly loquacious and outgoing. We
finally arrived at the Ostankino hotel in the northernmost reaches of
Moscow about 2 A.M.

Our room was virtually bare: no window curtains, two narrow
single beds sunken low in the middle, a bare bulb hanging from the
ceiling, and a small radio, which Lucille unplugged from the wall
immediately on the assumption that it was also a listening device. She
also searched behind the mirror in the bathroom for a hidden camera
or microphone. Indeed, we found ourselves speaking in whispers
throughout the week, having become paranoid as a result of a briefing
I received by a U.S. Air Force intelligence officer before leaving home.
The mid-June sunrise in Moscow was about 3:30 A.M., so that the sun
brightened our room shortly after we had fallen asleep. The next night
we pinned the bedspreads over the windows.

While I attended the meetings, Lucille joined five other wives and
three children from the United States, Canada, Japan, and Norway on
tours of Moscow under the leadership of June Yaglom. June is a
wonderful woman who was a freelance translator of English and
French literature to Russian at the time. She had never been out of the
USSR and yearned to see the original artwork that the foreign women
had seen in their travels. During the tours, Lucille had the distinct
impression that she was under surveillance because one of the "baby-
sitters" accompanied her everywhere, including her visits to the ladies
room. While resting one day on a park bench, she was approached by
a Russian man who asked her for the time in Russian, again arousing
her suspicion. This was about the time of *The Penkovsky Papers*—a
book dealing with Col. Penkovsky, who was a spy for the United States.
If memory serves me, he passed microfilms to the wife of a U.S.
military attaché while she walked her baby in the park.

Each morning the foreign delegates were bussed to the University
of Moscow on the south side of Moscow. One wonders why we were
housed an hour's ride on the other side of the city. One evening we and
the Gjessings had tickets to the opera at the Bolshoi Theater, and so we
managed to skip the bus back to the Ostankino and to escape from

Galia. It was a lovely experience, topped by dinner in the Aragvi Restaurant, the best non-Intourist restaurant in Moscow. But we had problems ordering dinner without Galia's translation. Luckily, however, the next table was occupied by the French delegates and their interpreter, so the ordering sequence went from English to French to Russian to the waiter who responded "Nyet" until we finally settled on chicken Kiev, which was available only in the good restaurants; ordinary Russians never saw it. It was just about midnight when we returned to Red Square to try to find a taxi to our hotel. (The famous Moscow subway stopped running then.) But the taxi shift was also changing at midnight and none of the drivers would take us. Again, we lucked out when the French delegation came along with their interpreter. When we explained our dilemma she spoke a few "persuasive" words to one of the drivers and he took us. Notably, none of the taxi drivers would accept tips from the despised foreign capitalists.

Yaglom recounts the joke about Yuri Dolgorukii, the founder of Moscow, whose monument is near the Aragvi: "How clever it was of Yuri Dolgorukii to found Moscow so close to the Aragvi restaurant."

One of the scientific tours was to the research station at Obninsk, where they had installed an impressive tower about 300 m in height for turbulence studies. Within the tower there was a tiny elevator capable of carrying three people. After the first group went up, a thunderstorm occurred and interrupted the power. There was no way that I was going to climb the ladder just to view the instruments at the lowest level—25 meters up. But two of the British scientists started the climb, followed by Georgii Golitsyn (who seemed in no physical condition to do it), and so I was impelled to hold up the glory of the United States, and followed them up. But without a guide the British guys missed the first and second platforms and climbed to the 75 meter level. That was a bit more than I bargained for. And just as I was taking a forbidden picture of the surroundings out stepped a Russian from the elevator, which was back in operation. He reprimanded me with restrained annoyance but did not confiscate my film.

After Moscow we traveled to Kiev (the city of my mother's birth), where I had an appointment to meet with Dr. Muchnik, one of the hydrologists who was using radar to measure rainfall. The Ukrainians in Kiev seemed a much happier and friendlier lot than the Muscovites, who always seemed grim, even though both were under the watchful eyes of "big brother." The contrast was evident in their faces and their colorful dress. My visit with Muchnik (and an unidentified baby-sitter) was very pleasant, but I barely survived the lunch of open sandwiches accompanied by abundant vodka and cognac. I suppose that I managed to hold my own because of my continued suspicion that they were trying to loosen my tongue. This was just one more event that kept us under stress throughout most of our stay.

We were relieved to board the KLM airplane for our trip to Amsterdam, to be greeted by the attractive and beautifully groomed flight attendants, and to receive a copy of the *International Herald Tribune.* But tension returned once more when a female Soviet officer tried to board the plane to remove an American physician who had worked there and was evidently leaving with rubles (a no-no). To his relief and ours, the aircraft captain insisted that the plane was Dutch territory and she had no right to be there. We all heaved a sigh of relief as the plane accelerated down the runway, and later cheered as we left Soviet airspace.

We returned to Moscow in 1971 for a meeting of UGGI that was attended by more than 2000 delegates. At that time we stayed in the new, centrally located, and massive Hotel Rossiya, with thousands of rooms. I have few recollections of the scientific sessions but only some of the social activities. These gigantic meetings are frequently short on substance.

We had maintained contact with the Yagloms in our annual New Year letter, and so we were invited to dinner at their apartment. Although he was not a member of the Academy of Sciences he had been allotted a larger apartment than the run-of-the-mill scientists, largely because he agreed to assume some administrative duties as head of laboratory in the Institute, and partly because a former student of Kolmogorov, who knew Yaglom well, had a high position in the governing body of the Academy. Nevertheless, the apartment was not spacious. Moreover, with their scientific and literary interests, there were books in every nook and cranny. Professor Lumley, from Cornell University, and a Russian scientist were also present. The dinner was excellent and the conversation surprisingly candid. Akiva discussed some of the difficulties he had encountered since 1965. Although he was still working at the Institute of Atmospheric Physics, life was not easy. It was a memorable evening that solidified our friendship.

I have only a vague memory of discussing immigration to the States. However, immigration to the United States was impossible then; his daughters began to dream about it much later and made the move in 1989. They were followed by June and Akiva in 1992.

During the week, Nikolai (Kolya) Vinnichenko, a turbulence expert from the Central Aerological Observatory (CAO), took us on a little tour of Moscow. Kolya had spent an academic year at Pennsylvania State University working with Hans Panofsky and Al Blackadar. During that period he also stayed with us during a brief visit to Chicago. (At that time he commented, "This house is just for one family?"). He was very pleasant, good company, and forthcoming in his opinions. He particularly enjoyed the *New York Times*. So it was a treat to have him give us a private tour of Moscow. When we passed the KGB building he joked that this was the tallest building in the USSR because one could "see" all the way to Siberia from there. At a church where we were not supposed to take pictures he suggested that I set the focus while he stood in front of me and take the shot when he stepped away. He also took us to the beautiful Novodevichje cemetery where Stalin's second wife (apparently a suicide) is buried alongside many other historic figures. The fact that Vinnichenko had a car at his disposal, a rare distinction at that time, surprised us but did not take on any deeper meaning until our return home, when one of my colleagues suggested that he was a member of the KGB. Similar rumors were circulating in Russia. I still find that difficult to believe, although Russian émigrés to the United States assure me that any Soviet citizen who was allowed to spend a year abroad in the 1960s was a member of the KGB.

One evening Kolya invited us to his apartment for a party. In addition to Lucille and myself, he had invited Ron Collis from Stanford Research Institute, Prof. Elmar Reitar from Colorado State University, and several of his colleagues from CAO. The one who stands out in my mind is Dr. Smirnova, who had worked with Andrei Gorelik on some of the early Doppler radar studies. (Her young daughter was an ice skater to whom we sent a skating skirt upon our return home.) The apartment consisted of a small living room that doubled as a bedroom, a hallway, a small kitchen, and a tiny toilet about the size of a phone booth. Ten or twelve of us were all gathered around a table set up in the living room. It was a festive occasion with abundant good food and much too much to drink. We toasted one another with vodka followed by brandy or cognac. We sang and danced in the minuscule space that remained open in the room. Near midnight, Kolya's wife brought out a plate of raw quail eggs, and "egged on" by our Russian friends, I participated in the custom of sucking the egg out. It did not sit well with the whiskey and I dashed to the toilet to throw up, but not fast enough to prevent soiling my only good suit. I spent the next half hour in the toilet while Lucille relayed wet rags to me so that I could retain some semblance of dignity. It was time to return to the hotel, but it took another hour to get a taxi. Thirty years later that party remains a memorable event. In spite of my accident, I felt closer to my Russian friends.

One evening we went to the main dining room in the hotel. The room was packed and we were about to leave when I caught the eye of Dr. G. K. Sulakvelidze from the High-Mountain Geophysical Institute (in the Caucasus), where he was one of the leaders of their hail suppression work. As soon as he saw us, he invited us to join his table. Wearing a star on his lapel that identified him as a recipient of some high honor, he quickly arranged for additional chairs and we joined them for dinner. He ordered wine and we spent more time drinking and toasting than eating. It was a pleasant evening but I suffered the effects of overindulgence with a sleepless night.

Sulakvelidze and associates had published an interesting treatise on the modification of hailstorms. Its central concept was the existence of "accumulation zones"of large supercooled raindrops that could be seeded by silver iodide to prevent growth to large-size hail. Oddly enough, this work was based in part on our earlier work in England on the detection of hail by dual wavelength radar. The accumulation-zone concept influenced the design of the National Hail Research Experiment (NHRE) that I was later to direct at the National Center for Atmospheric Research (Chapter 6). Indeed, the reports of successful hail suppression in the USSR were the key justification for the initiation of NHRE.

During one of the breaks in the 1971 conference I took a walk with Gorelik and Chernikov of CAO. Among other things, we spoke about the work of Sulakvelidze and I asked what they thought about it. They answered that "he was a nice man," which revealed much about their true feelings.

After my devastating experience with hail suppression (Chapter 6), I pulled no punches in addressing the Soviet delegation attending the conference on weather modification in Boulder in 1976. I told them that until they revealed their data on hail suppression the world would not accept their claims. It is notable that the hail suppression experiment (Grossversuch IV) that was conducted in Switzerland, following the detailed design of the Soviet methods, failed to confirm their results. Since then we have heard nothing further about hail suppression in the former Soviet Union.

I also met Prof. Kusiel Shifrin from Leningrad at the meetings. He and his wife had translated my 1964 monograph into Russian. It had been published in paperback and was for sale at 90 kopeks—about $1.00 at the official exchange rate, or 25 cents on the black market. The monograph was part of Volume 10 of *Advances in Geophysics,* which was selling at about $25 in the States at the time. I inquired about the possibility of collecting royalties—to little avail. (Lou Battan had received royalties on his first book on radar meteorology, but had to spend all his rubles before leaving the country.) I was pleased to be among those who helped Prof. Shifrin obtain an appointment at Oregon State University when he came to the United States at the end of the Cold War. Unfortunately, his wife died shortly after coming to the United States, but he continues there productively. I have also continued a friendly correspondence with Prof. Vladimir Stepanenko, of the Main Geophysical Observatory in St. Petersburg, to the present day. We had known each other for many years through exchanges of our papers on radar meteorology and cloud physics and his occasional visits to the United States as a member of a U.S.-Russian working group on environmental pollution.

I maintained contact with my Russian colleagues for some 15 years after my last visit and invited their participation in the 40th Anniversary Radar Meteorology Conference (Chapter 7). However, I received no response. The resulting book *Radar in Meteorology* is not complete without a chapter on the impressive work that had been going on in the USSR. It is unfortunate that the paranoia and xenophobia that characterized the Soviet bureaucracy during the Cold War prevented the cross-fertilization that is so important to the vigor of science. Their intransigence only served to hurt themselves.

At this writing it is difficult to conceive of the stressful relations between the United States and the USSR during the 45-plus years of the Cold War, and how that conflict affected us all personally. Incongruously, recent events demonstrate that two-directional spying is going on even while our astronauts work with the Russian cosmonauts on the international space station. But I am delighted to note that June and Akiva Yaglom and their children and grandchildren are living happily in Boston, and Valerian Tatarskii is working productively at the NOAA Environmental Technology Labs in Boulder. I have also enjoyed working and corresponding with two bright young Russian scientists, Sergey Matrosov in Boulder and Alexander Rhyzkov in Norman, Oklahoma.

CHAPTER 5

The University of Chicago—1966–1972

In view of the considerable institutional support that the Weather Radar Branch was receiving at AFCRL, the array of experimental tools at our disposal, and the flexibility we had to pursue our interests, it is perhaps surprising that I should have moved to the University of Chicago. A number of factors converged at the time to instigate the move. No one of these was decisive alone.

First of all, I had been at AFCRL for 18 years, and there was a natural instinct to try something else. This feeling was enhanced by the departure of a number of the senior members of the laboratory during the previous few years to form the Geophysics Corporation of America. A more immediate factor was the resignation of Roger Lhermitte to join the Sperry Corporation and then the NSSL, Keith Browning's return to England, and Ed Kessler's earlier departure to join Travelers Research Corporation and later to assume the directorship of the NSSL.

The early 1960s was also the time when I had reached some maturity and distinction as a scientist and started to receive offers from various universities. But none of these was sufficiently attractive academically, geographically, or financially. (I already had a supergrade civil service appointment so that most universities could not match my salary.) It was during the 1965 cloud physics meeting in Reno, Nevada that Paul MacCready invited Roscoe Braham and me to join him on a flight around the area, and Roscoe suggested that I consider moving to Chicago. He pursued the idea further in phone calls through early 1966. With a faculty that included Platzman, Fultz, Kuo, Hines, Braham, Pedlosky, and Fujita, and the distinction that the University of Chicago enjoyed across the board, it was a most attractive place. But the physical climate was not any better than that in Boston. So I declined, but agreed to come out to present a seminar. This was at the time that we were getting the revolutionary observations on clear-air echoes at Wallops Island, so it was not difficult to excite the interest of the audience. As I discovered later, the audience included most of the 25 or 26 members of the faculty of the Department of

Geophysical Sciences—geologists, paleontologists, and meteorologists. Immediately following the seminar, Prof. Julian Goldsmith, chairman of the department, offered me a tenured full professorship at a most attractive salary. I did not accept then, but I was sufficiently starry-eyed and flattered to go home in a very receptive mood.

Since I had great ambitions to build a major experimental radar facility, I subsequently returned to Chicago and requested a meeting with President George Beadle (a Nobel laureate) to make sure that the university understood my plans and would support them. Beadle was a delightful guy (who insisted that I call him "George") and expressed enthusiasm for everything I proposed, although it was clear that I had to bring in the funds. Julian and Ethel Goldsmith entertained us graciously, and Ethel showed us around Chicago and the suburbs to check out housing. We were overwhelmed, and I accepted the offer. I must also admit to being attracted to the proximity to Lake Michigan, where we could water-ski.

The University of Chicago is located on the south side of Chicago. While the surrounding neighborhoods are depressed, the immediate community had and has its own distinctive charm—with a mix of homes dating from the early twentieth century, modern high-rises, and town houses. At the time of our stay, roughly 70 percent of the faculty lived within walking distance of the campus, making the neighborhood a small town. One would meet friends at the supermarket or on the street every day. The Department of Geophysical Sciences (DOGS) was a particularly friendly group, and socialized frequently. Julian Goldsmith was a high-spirited, good-humored, and hospitable chairman who set the tone for both the intellectual and social life of the department. We quickly made friends, were invited to a variety of social events, entertained our friends, and felt very much at home in short order.

My immediate tasks were to teach a course in radar meteorology in the fall of 1966 and to seek funds for our research. In anticipation of my joining the university, I had prepared a proposal to the NSF calling for the development of a major Doppler radar built around a 10.7 cm FPS-18 system. One of the big challenges was the design and development (under contract) of a 25-ft diameter antenna and pedestal that could be readily disassembled and moved. Prior to the invention of the pulse pair processor by Rummler at Bell Telephone Laboratories in 1968, the development of a signal processor was also a formidable task.

In 1967, we received about $650,000 from the NSF for this job. That sum was the total NSF budget for meteorological instrumentation in the nation. It was an expression of the high hopes that the NSF had for the future of Doppler radar and the expectation that we at Chicago would develop a strong program in radar meteorology. Robert Serafin, then a staff member at the Illinois Institute of Technology (IIT) Research Foundation and a doctoral candidate at IIT, was the key project engineer on the system design. Later, we also developed a cooperative arrangement with the Illinois State Water Survey (Eugene Mueller and Don Staggs) for the installation and operation of the radar. It was this collaboration that produced the name CHILL, for CHicago-ILLinois radar. After many upgrades, CHILL has become the modern multiparameter radar now operated as a national facility under the aegis of Colorado State University.

Since the University of Chicago did not have an engineering school, we formed a Joint Laboratory for Atmospheric Probing with the Department of Electrical Engineering at IIT. We thus gained the collaboration of professors Lester Peach and Ken Gage, and of Bob Serafin. In addition to his responsibilities for the development of the CHILL radar, Bob did his doctoral dissertation on the measurement of atmospheric turbulence under the joint guidance of Prof. Peach and myself. Several other IIT students also did their theses on radar meteorological techniques. The dissertation of Dennenberg (70) on the measurement of spectral moments was especially important. We also installed a powerful vertically pointing Doppler radar with a 40-ft diameter dish on the roof at IIT.

A significant contribution to our startup was a modest grant by United Airlines (UAL) for a period of five years. This grant was based largely on my earlier developments of radar storm contour mapping and airborne weather radar, first used operationally by UAL, and our subsequent work on the detection of CAT. Boynton Beckwith, director of meteorology for UAL and the leader of UAL's initial experiments in airborne weather radar in the early 1950s, must have had a strong influence on the UAL grant. Because the grant was unrestricted, we could use it in any way necessary to leverage the rest of the program. The best investment I made was to use those funds to bring Dr. Ramesh Srivastava from India to the University of Chicago as a research associate. He quickly became an important member of the faculty and remains there until this day. More will be said about him later.

As was the case in our prior activities, we did not wait for the completion of our own radars but sought access to existing data and experimental facilities elsewhere. In 1967, I spent much of the summer as a guest of Prof. Birkemeier at the University of Wisconsin. He and his students were conducting forward-scatter beam-swinging experiments using a 900-MHz radio link between Madison and Cedar Rapids, Iowa. However, they were having difficulty explaining the behavior of the signal strength and the Doppler shift as a function of the deviation of the beam angle from the great circle path. Since I had recently been involved with the clear-air backscatter studies at Wallops Island, where the turbulent scatter was clearly confined to thin layers, and been thoroughly grounded in the turbulent scatter theory of Tatarski (35) as well as Doppler measurements of the winds, it was possible for me to put two and two together in short order. The result was an exciting set of papers in association with Dr. Srivastava and Chicago students Richard Carbone and Walter Marker and Wisconsin student Douglas Sargeant (71, 72). Sargeant was later to play a major role in the Global Atmospheric Research Project (GARP) Atlantic Tropical Experiment (GATE).

The following year we were grateful to receive a set of Doppler radar observations taken during stratiform rain by Keith Browning, then in charge of the meteorological unit at the Royal Radar Establishment in England. This led to an interesting study of the cooling of the air by melting snow and the adjustment of the wind by the associated pressure perturbations, again with Srivastava, students Marker and Carbone, and visiting scientist Ryozo Tatehira, later to become the director general of the Japanese Meteorological Agency (73).

Perhaps the most thrilling experiments were those conducted as the guests of Dr. Juergen Richter at the Naval Electronics Laboratory Center, at Point Loma, San Diego. Just prior to the June 1969 meeting in Stockholm of the Inter-Union Committee on Radio Meteorology on the spectra of meteorological variables, I received a set of unique observations from Richter. These were taken with a first-of-its-kind frequency modulated/continuous wave (FM/CW) zenith-pointing radar that had both ultrahigh sensitivity and unprecedented range resolution of only 1 meter. This compares to typical weather radar resolutions of 150 m. These observations showed exceedingly fine clear-air echo layers, frequently associated with breaking waves, located exactly at well-defined inversions and sharp changes in the vertical gradient of refractive index. Some of these layers were no thicker than the 1-meter resolution. Having seen similar echo layers with the powerful radars at Wallops Island with much poorer vertical resolution, I could hardly believe my eyes. Galileo, when he first observed the planets by a telescope, could not have been more excited than we were. But I recalled the observations of the coherent echoes that we had detected from the sea-breeze inversion (30) with our 1.25 cm radar. The latter had implied refractive index gradients of $\approx 4\,N$ units (parts per million) per centimeter. Here, for the first time, were ultrahigh-resolution radar observations that suggested the same order of magnitude.

With Richter's encouragement, I presented his observations at the Stockholm meeting. At first, they were received with astonishment, but the observations were beyond question, and so they quickly became part of the folklore of clear-air scatter. Upon my return to Chicago, I departed for San Diego to spend a month with Richter, exhilarating over ever more exciting observations of the ultrafine structure of the atmosphere. The following summer, I returned with graduate student James Metcalf and postdoctoral associate Ernst Stratmann. This time we had made prior arrangements with Don Lenschow of NCAR and James Telford of the Desert Research Institute, University of Nevada to use the Buffalo airplane with its inertial navigation and gust probe measurement systems. This was their first use in a research experiment and they worked magnificently, largely with the sweat of Jim Telford to nurture them daily. The results were beyond our fondest dreams, with papers on the birth of CAT and microscale turbulence (74) and internal waves in the atmosphere (75). And two years later, there was the paper on the remarkably well-ordered microscale structure of the Kelvin-Helmholtz waves of only 10-m amplitude and 80-m length revealed by both the aircraft and radar measurements (76). The latter work was central to Metcalf's doctoral dissertation.

Meanwhile, a very active research program was under way in Chicago. Sekhon and Srivastava did basic studies of the reflectivity of snow particle size spectra (77) and raindrop size spectra in a thunderstorm as deduced from Doppler radar observations (78). Interesting work was also done on the radar detection of hail in a series of papers by Eccles, Carbone, Srivastava, Jameson, and myself. Hildebrand and

Sekhon (79) did a fine study on the objective determination of the noise level of Doppler spectra. This was Hildebrand's first exposure to Doppler signal analysis, which was later to stand him in good stead when he led the development of the highly sophisticated airborne Doppler radar (ELDORA) at NCAR (80). An excellent theoretical study was also done by Ekpenyong and Srivastava on the radar characteristics of the melting layer (81). Harris then followed this with a comprehensive Doppler radar study of evaporating ice clouds (82). Andrew Heymsfield, working under Prof. Braham, collaborated with our group in his dissertation on the nature of cirrus uncinus (mare's tail) clouds, which has become a classic in the field (83). And Bob Serafin did an impressive paper on the radar measurement of turbulence with Prof. Peach and myself (84). This later became his dissertation at IIT and resulted in a U.S. patent.

When it came to precipitation physics, Srivastava was a most prolific leader and contributor. He undertook studies of the growth and motion of precipitation particles in convective storms (85), the effect of collision, coalescence, and breakup on drop-size distributions (86), and the above-mentioned work on the melting layer, evaporation, and snow- and raindrop size spectra. As an extension of our coursework, we also collaborated on a comprehensive review paper on the Doppler radar characteristics of precipitation at vertical incidence, which has become a basic reference in the field (87). Needless to say, much more was done under Srivastava's leadership subsequent to my departure from the University of Chicago. Indeed, except for Metcalf and Serafin, it was he who guided the rest of the students through to their doctorates, although I participated as a member of their committees.

Although incomplete, the above account serves to indicate the range and intensity of the research accomplished by the students, postdocs, and associates of the Joint Laboratory for Atmospheric Probing at the University of Chicago and IIT. In retrospect, it was an impressive record of achievement. Perhaps its significance can best be judged by the subsequent attainments of the students who went on to high distinction in their own right at other institutions. I can only speculate that the apparent creativity of the group both at Chicago and in their later positions was due in part to their in-depth exposure to both the radar methodology and the physics of the atmosphere, and partly to the exciting environment of discovery. We must also acknowledge the general atmosphere of intellectual momentum that characterized the department and the university. It is noteworthy that Bob Serafin later followed me as director of the Atmospheric Technology Division at NCAR, and Rit Carbone followed him when Bob moved up to become the NCAR director.

It was also a joy to interact with the students socially. Another professor and I played volleyball with the graduate students once a week, but I gave up tennis with Rit Carbone after he beat me badly several times. I also enjoyed skiing with Ryozo Tatehira and Peter Black in southern Wisconsin. And Lucille and I entertained them occasionally at home. Carbone surprised me on my first birthday at the university when he organized a party and presented me with a child's science kit titled "Prof. Wonderful." And many of the students shared brown-bag lunches with Srivastava and me most days during which the subject matter ranged widely. It was particularly interesting to hear them express ideas on how they expected their lives to evolve along nonconventional lines. But most have followed the practices of my own generation—marriage, children, careers, and concerns about raising and supporting children through college. I am still in frequent contact with these "young" students—who are now in their mid- to late fifties—and enjoy their friendship.

We come now to teaching—which turned out to be a daunting experience. When I accepted the position, I simply assumed that I could lecture to students as I had done at scientific meetings—with clarity and enthusiasm. But conference halls and classrooms are at different ends of the teaching spectrum. In the former, one assumes that the audience knows the fundamentals and cuts right to the chase—the goal, the experiment, the observations, discussion, and conclusion, all in 10 to 15 minutes, and preferably without distracting equations. In the classroom, however, fundamentals come first. And I had long since forgotten the theoretical foundations of much of the subject matter, nor had I ever studied some that were necessary. So, it was a major undertaking to relearn electromagnetic theory, and to learn stochastic processes and signal theory, which are the bases for understanding the backscatter from individual particles and random arrays of raindrops and ice crystals. My thermodynamics was rusty and my dynamics was nil. Nor did I have a grounding in turbulence theory and measurement. When I realized all that I was lacking, I was sure that they had made a mistake in hiring me.

So, I buckled down to study. During the first year, I struggled through, using some elementary texts in electromagnetic theory, information theory, and the basic papers by my colleagues at McGill University and MIT. I also relied heavily on my 1964 monograph, which reviewed much of what was known about radar meteorology at the time. I also depended on the excellent papers of Roger Lhermitte, and the superb dissertation of Rod Rogers on Doppler radar. By the second year, Ramesh Srivastava arrived and eased my burden. We collaborated on developing a wonderful set of course notes that amounted to some 13 loose-leaf volumes by 1970.

Ramesh quickly learned all that was necessary in radar, signal theory and measurement, and the manner in which the atmosphere interacted with the radar to provide useful information about the physical structure and processes. I am an intuitive thinker, and could make an educated guess to interpret the radar observations in terms of the physical processes. But Ramesh could translate the measurements to physical mechanisms in a rigorous mathematical manner. We complemented each other well. He taught me a great deal in those six years. And of course, as is usually the case, the professor learns more from the teaching process than the students do. I also learned much from my colleagues in electrical engineering at IIT, especially from Bob Serafin, who was an active member of the joint laboratory.

Ramesh Srivastava is one of the most versatile scientists I have known. There was no problem in precipitation physics, radar meteorology, or other topics in meteorology that he could not tackle and quickly undertake seminal research. He is universally respected but has not garnered high awards because of his self-effacing character. I regret not having nominated him for such honors.

We had developed such a nice set of course notes by 1970 that we submitted a proposal to the University of Chicago Press to publish a book on radar meteorology. However, I was a member of the board of editors of the University of Chicago Press at the time, and we received a revised edition of Lou Battan's first book on the subject, which I was asked to review. Although his book was not as comprehensive as that which we had planned, it was a fine volume, and it was ready. Ours would have taken another year or more to complete. So, I resolved my ambivalent feelings by recommending its publication, recognizing that his book would preempt ours.

As a sidelight, my two-year membership on the board of editors was one of the highlights of my University of Chicago tenure. The board was comprised of distinguished faculty representing many of the disciplines. I was honored to be among them. Each month we would receive a thick package of reviews of some ten or more books. Many of the reviews were so comprehensive that it was an education to read them. Although my understanding of the subject matter was usually limited, I gleaned enough from the reviews and the discussions to participate in the decisions.

In addition to the traditional university activities, there were also some appealing extracurricular endeavors. The first of these was a joint consulting task that Prof. Ted Fujita and I undertook on behalf of a large discount store damaged by a 1967 tornado. In spite of the striking damage to the store, the insurance company refused to pay for the apparent damage to the roof rafters. The tornado first touched down in Oak Lawn, killing more than 50 people, and wreaking great damage. Thereafter it lifted, moved eastward, and touched down in an open field just west of the store at Cottage Grove Avenue on the south side of the city. Although our investigation of the damage took place two years after the event, it was clear that the tornado had hit the store. This was evident from the window breakage; the windows on the east side of the northeast corner of the building were blown outward, while those on the west side, just 30 to 50 ft away, were blown inward, thus indicating the cyclonic rotation. Incredibly, witnesses reported that a sheet of glass had cut through the metal hood of a parked car. In spite of this evidence, the insurance company doubted that the damage to the 12 by 12 inch roof rafters had been done by the tornado. During further inquiries on Cottage Grove Avenue, we found the owner of a restaurant who watched his Pontiac automobile being lifted off the street and carried some 60 feet across the way where it was dropped on another parked car. We were able to get the lift coefficients for the car and estimated the range of wind speeds required to lift it. We were then able to use these estimates to compute reasonable lift forces on the curved roof of the store. The insurance for the roof damage was paid. This episode was an example of Ted Fujita's uncanny ability to infer the structure and intensity of tornadoes and microbursts from the nature of surface damage.

Another notable experience was the case of a United Airlines pilot who was fired by the airline for pilot error and had his license revoked after a curious accident. I was retained by the Airline Pilots Association to assist in his defense. (It was a bit awkward for me to take a position opposing my benefactor Boynton Beckwith and UAL.) The aircraft had taken off from Detroit Metropolitan Airport in a light rain and was cleared to 4,000 ft altitude. The pilot was then cleared to 22,000 ft, but before he applied power, the plane was mysteriously carried up to 17,000 ft. The pilot then nosed over to escape the apparent updraft, when he went into a dive at about 30,000 ft per minute. In the nick of time, he pulled up at about 500 ft above the ground. He was then able to land the plane safely at Detroit without injury to any of the passengers. In subsequent interviews of the passengers by the civil aviation authorities, one of the passengers (a meteorologist and pilot) expressed the view that the severe undulations were pilot-induced. This was evidently the proximal cause of the loss of license, although I believe that he had been involved in another prior incident. Coincidentally, I had recently returned from San Diego where I had observed some rather large amplitude breaking waves and clear-air turbulence, and similar waves had been observed regularly at Wallops Island. With these observations and some supporting surface barometric data showing sharp perturbations in surface pressure at the time of the incident, the pilot was cleared and regained his license and job. Admittedly, our evidence was strictly circumstantial, but it sufficed to convince the "master" (acting in lieu of a judge or jury) that it was not pilot error.

Between 1967 and 1969, I also chaired the panel on remote atmospheric probing of the committee on atmospheric sciences of the National Academy of Sciences. In view of the load that I was already carrying at the University of Chicago, it was probably a foolhardy decision to accept this responsibility. But I did so because of my view that the panel's goal was important to the future of the science. The panel had several meetings during which we heard presentations on the state of the art and future potential by many of the nation's experts on the subjects. Our final report (88) included recommendations on: (1) the expanded use of ultrasensitive radar for probing the clear atmosphere; (2) the vigorous development of single and multiple station Doppler radar and airborne Doppler radar for studies of atmospheric motions and tornado detection; (3) the investigation of the potential of acoustic techniques in remote probing; (4) the active pursuit of lidar along a number of specific directions; (5) the application of microwave radiometry for profiling atmospheric temperature and water vapor from both the ground and space, and mapping precipitation from space; (6) further experiments on the use of spaceborne infrared sensors for the profiling of temperature, water vapor, and ozone; and (7) the use of microwave and optical line-of-sight paths to infer atmospheric structure, content, and motion.

Some of these techniques had already been investigated and simply required the blessings of our panel, while others were newborn and judged to have considerable potential. Acoustic radar (SODAR—for Sound Detection and Ranging) was the most novel approach at that time, and soon became a valuable tool for probing the boundary layer. This was due in large part to Gordon Little, director of the wave propagation lab of NOAA Environmental Research Labs, who took the initiative and later made it a topic of his own research. In contrast, Stewart Marshall, who had pioneered radar meteorology, surprised us by not jumping on the Doppler radar bandwagon. The result was that the Stormy Weather Group at McGill University delayed moving into Doppler studies for more than a decade.

One cannot assess the impact that the panel report had on the evolution of remote probing. Probably its most important contribution was that investigators could refer to the document to support their proposals and that program managers at the funding agencies could use it to support their funding decisions. It is noteworthy that most of the techniques have been successfully pursued in one way or another; some have led to remarkable discoveries, and others have become operational.

By mid-1971, the pace I was leading had worn me to a frazzle. In order to operate with a total staff of 12 to 15, including five or six graduate students, two or three postdocs, a University of Chicago faculty member, and the engineers and technicians at IIT, we had to bring in about $350,000 per year, a major sum at the time. (Before I accepted the position at the University of Chicago, Lou Battan had suggested that I could succeed if I could manage with a grant of about $100,000 annually; he was right.) And I had to do it by myself; my colleagues at IIT had not been able to acquire independent funding. But I could only do it with the support from four agencies—the NSF, the Environmental Protection Agency, the Office of Naval Research, and the U.S. Army Research Office. (The bookkeeping became so complicated that our secretary refused to do it, thus leaving the task to my wife, Lucille.) So, proposal writing and selling was a major part of my job. It was a superhuman task, and I was no superman. I started to yearn for the good old days at AFCRL when the funds flowed more readily. And I began to speculate about how I could ease my burdens. After confidential discussions with friends and colleagues, the word on the grapevine was out that I was movable, and I received several invitations. Among the most attractive of these was that from Walter Orr Roberts and John Firor at NCAR to assume the directorship of the Facilities Division. After much discussion with Lucille, I accepted, subject to the completion of the academic year and the transfer of at least one of my grants to Srivastava.

I felt like a traitor to abandon the University of Chicago, the students who were still working toward their advanced degrees, and my colleagues in the department. Friends like Julian Goldsmith were Chicago chauvinists, so there was no way that I could justify leaving the university in their eyes. Nor did we cherish the thought of leaving the many friends we had made and starting over again in Boulder. In addition, our daughter Joan was just finishing her bachelor's degree at the University of Rochester, with tuition paid by the University of Chicago, and our son was starting at Oberlin College under a similar arrangement. This only served to heighten my guilty conscience about departing Chicago. However, I did not burn my bridges; I took leave of absence.

Illustrations

Plate 1. Lt. David Atlas at All Weather Flying Division, Clinton County Army Air Base, Wilmington, Ohio, 1946. The belly dome on the B-25 weather research airplane contains the antenna of the AN/APQ-13 radar.

Plate 2. Louis Battan, the best man, David Atlas, and Robert Siegel at Dave's marriage to Lucille on September 26, 1948, Riverside Plaza Hotel, New York City.

Plate 3. Members and associates of the Weather Radar Branch on Great Blue Hill, Milton, Massachusetts, 1957. Left to right, back row: Graham Armstrong, Edwin Kessler, Mary Weaver (Harvard Blue Hill Observatory), Stanley Yorkin (HBHO), Kenneth Ultsch (HBHO); middle row: David Atlas, William Lamkin, Albert Chmela; front row: Ruben Novack, Ralph Donaldson, Robert Whittaker. Photo by John Conover. Copyright: American Meteorological Society.

Plate 4. David Atlas, Raymond Wexler, Bob Fletcher (AMS president and Chief Scientist, Air Weather Service), and Col. Bob Houghton (Air Weather Service) at Seventh Weather Radar Conference, Miami Beach, Florida, November, 1958.

Plate 5. Left to right: David Atlas, Vaughn Rockney (head of Weather Radar Program, US Weather Bureau), Mrs. and Dr. Morris Tepper, Ron Collis, Lucille Atlas, Dr. and Mrs. Richard Douglas. Seventh Weather Radar Conference, Miami Beach, Florida, November, 1958.

Plate 6. Rear: Professor Nelson Dingle (University of Michigan) and David Atlas; near side of table, left to right: Mrs. Jean Hiser and Homer Hiser (University of Miami), Mrs. Geotis and Spiros (Speed) Geotis (MIT Weather Radar Project); far side of table: Walter Hitschfeld (McGill University), Mel Stone and Mrs. Stone (MIT), Mrs. and Robert Copeland (MIT). Seventh Weather Radar Conference, Miami Beach, Florida, November, 1958.

Plate 7. Left to right: David Atlas, Rilda and Stanley Mossop (CSIRO, Australia), Frank Ludlam and son Hugo (Imperial College), Mrs. Jean Ludlam, William Macklin and Keith Browning (graduate students), and unidentified couple at Ashridge House, Little Gaddesden, England, during the Wokingham Storm project, June 1959.

Plate 8. Roger Lhermitte (Emeritus, Univ. of Miami), Gen. Benjamin Holtzman (Commander, AFCRL), and David Atlas when we were recognized for the patent on the Doppler radar measurement of winds by the velocity-azimuth display (VAD). AFCRL, 1963.

Plate 9. Past presidents of the American Meteorological Society at the national meeting in Philadelphia, January 1976. Left to right, back row: Alfred Blackadar—1971; Phillip Thompson—1964–65; Louis Battan—1966–67; David Johnson—1974; William Kellogg—1973; David Atlas—1975; front row: Morris Neiburger—1962–63; Bernard Haurwitz—1943; Francis Reichelderfer—1940–41; Charles Hosler—1976; Henry Houghton—1946–47; Thomas Malone—1960–61.

Plate 10. At the retirement party of Prof. Stewart Marshall, McGill University, 1979. Left to right: Walter McKenzie Palmer, Stewart Marshall, Walter Hitschfeld, and David Atlas.

Plate 11. David Atlas, Richard Hallgren (AMS president), and Joanne Simpson when Dave received the Cleveland Abbe Award and Joanne received the Carl Gustav Rossby Award at the AMS national meeting in January, 1983.

Plate 12. David Atlas and Robert Serafin (Director, National Center for Atmospheric Research) at NCAR, 1988.

Plate 13. Col. Marcellus Duffy (U.S. Air Force, Retired), one of the key Army Air Corps officers who encouraged the training of weather officers in radar during WW II. Pauline Austin, head of the MIT Weather Radar Project for many years, and David Atlas. Taken at the 40th Anniversary Radar Meteorology and Battan Memorial Conference in Boston, November 1987.

Plate 14. My colleagues, Ralph Donaldson and Albert Chmela who cajoled and stimulated me during the early days at Air Force Cambridge Research Laboratories, and Stuart Bigler, retired former head of the National Weather Service Weather Radar Program and Officer in Charge of the Anchorage, Alaska Weather Forecast Office.

Plate 15. Three of the old MIT weather radar gang. L to R.—Prof. Reggie Newell, Prof. Rod Rogers, one of the pioneers of Doppler Radar, then at McGill University, and everyone's favorite, Spiros (Speed) Geotis, the long time radar engineer for the MIT Weather Radar Project. Taken at the 40th Anniversary Radar Meteorology and Battan Memorial Conference in Boston, November 1987.

Plate 16. L–R: Dr. John Freeman, one of the pioneers of numerical prediction and a long-time devotee to radar meteorology; Dr. Nobuhiko Kodaira, who did the pioneering work on signal integration by a delay line integrator at MIT and a leader of the research in radar meteorology at the Meteorological Research Institute in Japan; and Isadore Katz, who conducted the seminal research in propagation of radar waves at the MIT Radiation Laboratories during WW II. He also headed the radar group from the Johns Hopkins Applied Physics Laboratory that collaborated with us at Wallops Island in the studies of clear air echoes and the detection of clear air turbulence. Taken at the 40th Anniversary Radar Meteorology and Battan Memorial Conference in Boston, November 1987.

Plate 17. L–R: Ken Spengler, Executive Director of the AMS. Mrs. Jeannette Battan, holding the collected works of her husband, Lou, which were presented by Phillip Krider, Director of the Institute of Atmospheric Physics, University of Arizona. Taken at the 40th Anniversary Radar Meteorology and Battan Memorial Conference in Boston, November 1987.

Plate 18. Prof. Roscoe Braham, University of Chicago (then President of the AMS) presenting the Jules Charney Award of the AMS to Dr. Keith Browning, then Director of Research at the U.K. Meteorological Office. Taken at the 40th Anniversary Radar Meteorology and Battan Memorial Conference in Boston, November 1987.

Plate 19. L–R: Prof. Richard Douglas, an early member of the Stormy Weather Group at McGill University; Doug Jones, an early member of the weather radar and rainfall studies group at the Illinois State Water Survey, and Edwin Kessler, a colleague of mine at the Weather Radar Branch, AFCRL, and Director of the National Severe Storms Laboratory at Norman, Oklahoma. Taken at the 40th Anniversary Radar Meteorology and Battan Memorial Conference in Boston, November 1987.

Plate 20. Reunion of former students in radar meteorology, University of Chicago. Left to right: Rit Carbone, Peter Hildebrand, David Johnson, Andy Heymsfield, John McCarthy, Bob Serafin, and David Atlas, Boulder, Colorado, 1994.

Plate 21. Left: David Atlas, founding director of Goddard Laboratory for Atmospheric Sciences, NASA; right: Marvin Geller, second director (now Laboratory for Atmospheres), and Franco Einaudi, most recent director, at the 50th anniversary celebration of the career of Joanne Simpson, December 1999.

Plate 22. Except for David Johnson then AMS President and leader of our delegation, this is the AMS delegation that went to the People's Republic of China in April 1974. Left to right: Joan and Dick Reed, Betty and Will Kellogg, Peg and Ken Spengler, Lucille and Dave Atlas. The photo was taken on Jan 12, 2001 at the 81st annual meeting of the AMS in Albuquerque, New Mexico.

Plate 23. Here are shown some of my colleagues with whom I have enjoyed remarkable symbiotic interactions over many years. Left to right—Dr. Robert Meneghini, with whom I have worked closely since joining NASA in 1977 and one of the leaders of the TRMM satellite program; Prof. Carlton Ulbrich of Clemson University, with whom I have collaborated regularly since 1972 at NCAR and NASA; Dr. Toshio Oguchi, Communications Research Laboratory, Japan, a key contributor to the TRMM satellite program, Dr. Frank Marks of the Hurricane Research Division, NOAA Atlantic Oceano-graphic and Meteorological Laboratories, Miami, with whom we did studies of hurricane and tropical rainfall; and on the far right, Prof. Daniel Rosenfeld of Hebrew University, Israel, who spent several exciting years working with us at Goddard on radar rainfall measuring techniques. Photo taken at the meeting of the TRMM Science Team at Greenbelt, Maryland, October 2000.

CHAPTER 6

National Center for Atmospheric Research—1972–1977

Section 6.1. Atmospheric Technology Division, 1972–1974

We arrived in Boulder in late August of 1972 and moved into an apartment at Remington Post at 30th St. and the "Diagonal" to Longmont in early September until we could decide whether or not to remain permanently. My new office was located at 30th and Arapaho so that it was a pleasant bike ride each way. I was to replace Dan Rex. My deputies were Charlie Palmer and Harry Vaughan.

The year 1972 was a painful one for UCAR and NCAR. The universities were suffering from budget cuts, and many faculty members were not getting the kind of support to which they had been accustomed. They were also unhappy that many NCAR scientists were doing "small" research jobs that could have been done by individual scientists rather than the "big" projects for which NCAR was supposedly designed. In general, they considered NCAR science to be competitive with that at the universities. And there was a general displeasure with the nature of the support coming from the Facilities Division, except that for the Computing Facility. In general, the universities looked upon NCAR as Camelot, the beautiful palace on the mesa where the scientists did as they pleased.

The trustees had appointed the Joint Evaluation Committee (JEC) to investigate the situation. By coincidence, I was in Washington at the time the report was ready to be delivered, so I was asked to pick up the box of reports at Johns Hopkins University. Thus, I was the inadvertent bearer of bad news. The JEC report was a strong indictment of NCAR science and management. As president of UCAR, Walt Roberts took most of the heat, although neither John Firor nor the division heads escaped severe criticism. To my surprise, the report referred to my appointment as a sign of hope for the improvement of the facilities. A comprehensive report of the upheaval at NCAR appeared in *Science* (Vol. 82, Oct 5, 1973, pp 36–39).

Ordinarily, a devastating report of this kind would mean the resignation of the president. However, Walt Roberts was so well liked that the trustees hesitated to take such severe action. Instead, they proposed the appointment of a strong manager as vice president. The work of the search committee (of which I was a member) was pre-empted when the trustees took action to ask for Walt's resignation. Subsequently, Prof. Francis Bretherton of Johns Hopkins University was appointed president of UCAR and director of NCAR.

My charge at the Facilities Division was to clean house and build it into a reputable operation. I was given carte blanche. Because of my vision that this division should be developing the instruments, plat-forms, and methods for observing the atmosphere, one of the first things I did was to change the name to the Atmospheric Technology Division (ATD) to reflect its more elevated mission. (Vincent Lally, famous for his clever acronyms, mischievously suggested that it be named "Advanced Technology Laboratory for Atmospheric Sciences"—or ATLAS).

Because NCAR had staffed up rather rapidly after its establish-ment in 1960, several of its senior staff were high-level military retirees who were no longer up to date on the science and lacked the entrepreneurial spirit of the more recently trained scientists and engineers. These included the heads of the Facilities Division, the Field Observing Facility, and the Aviation Facility. Indeed, except for the Computing Facility and the National Balloon Facility at Palestine, Texas, none of them was performing at acceptable levels. So I pro-ceeded with due deliberation to seek more dynamic and technologically qualified managers. I brought in Bob Serafin to head the Field Observ-ing Facility (later renamed the Remote Sensing Facility and subse-quently managed by Richard [Rit] Carbone). I also brought in David Bargen to head the Research Systems Facility, and later still, I re-placed the manager of the Computing Facility with Stuart Patterson. However, although I was later criticized for not replacing the manager of the Aviation Facility, I kept him on because he had very effective relations with the Navy and Air Force and could obtain aircraft and spare parts at little or no cost. This was no small contribution, because we were always operating on a tight budget. However, I did arrange a joint appointment for Don Lenschow, a scientist from the Laboratory for Atmospheric Sciences, to bring science know-how and objectives to the Aviation Facility. I paid little attention to the Balloon Facility because they were performing very well in support of the high-altitude research community. At first, the scientific staff in the other NCAR divisions regarded the scientific mission of ATD as competitive, but they soon found the improved performance beneficial to their own work.

It was gratifying to see the transformation that took place in ATD. The Field Observing Facility (FOF) developed a pulse-pair Doppler processor by copying that which had been built by Roger Lhermitte at the University of Miami. This processor revolutionized radar technology by permitting Doppler measurements to be made in real time at all ranges. They also developed the displays to present the vast quantity of radar reflectivity and velocity information in living color so that one could interpret the storm structure and flow patterns quickly and with little ambiguity. These were installed on the CP-2 radar and used very effectively by the National Hail Research Experiment (NHRE). (Note that the CP-2 was essentially a copy of the CHILL radar developed at the University of Chicago.) FOF also began the development of two transportable 5.6 cm wavelength Doppler radars (the CP-3 and CP-4) that could be used in the dual-Doppler mode envisioned by Lhermitte to get the full velocity vector instead of only the radial component measurable with a single Doppler radar. These radars were of inestimable value to the academic community for more than a decade. FOF also joined forces with the Research Systems Facility in developing the portable automated mesonet (PAM) under the leadership of Fred Brock, who came from the University of Oklahoma. PAM was a network of remote weather instruments that transmitted their data to a central base station so that the surface weather could be displayed in real time.

The Aviation Facility developed improved gust probes and other sensors, and advanced data processing and display systems. And Vince Lally's group continued to make impressive developments of the long-lived globe-girdling constant-level balloons and carrier balloons equipped to release dropsondes. Although the balloons were intended for measuring the winds in the Tropics during GARP, they could not be relied upon to remain in the tropical latitudes where winds were required. Thus, the balloons were abandoned in favor of reconnaissance aircraft during the special observing periods of the First Global GARP Experiment (FGGE). Nevertheless, constant-level balloons were later used to gain important insights into the global circulation. One of the balloons circled the globe for over two years.

Having had no experience as a manager of such a large group (comprised of some 300 people), I took a short course in management offered by the University of Colorado. I then introduced the system of "management by objectives" so that the engineers outlined their goals for the year and could be evaluated against their targets. This worked well in ATD but failed when I mistakenly tried to do the same in the National Hail Research Experiment (NHRE), where the scientists insisted upon the kind of flexibility that I had always advocated in my prior positions. I also established an internal advisory committee made up of ATD staff members to exchange ideas on morale, personnel, and NCAR practices. Since the ATD engineers and support staff were regarded as second-class citizens by the rest of NCAR, I suggested establishing awards for technological advances and research support, both of which were adopted and supported by Walt Roberts out of his personal funds. My impression is that we achieved a major improvement in the technical performance of ATD and its support to the university community while greatly enhancing staff morale. I was satisfied and happy.

During this period, Prof. Carlton Ulbrich of Clemson University spent a year as a visiting scientist under the advanced study program at NCAR. As a result of our mutual interests, we established a close working relationship and did some interesting research on the physical basis of the relationship between rainfall and its effect on the attenuation of microwave radiation (89). This was the beginning of a productive collaboration and a close friendship that has continued to the present time. Later I invited him to join us at NHRE and at NASA.

Section 6.2. The National Hail Research Experiment (NHRE), 1974–75

Dr. William (Bill) Swinbank, director of NHRE, died suddenly during Christmas of 1973. Bill was already an eminent scientist in turbulence theory and experiment. He had started his career in England and then moved to Australia before coming to the United States. I loved and admired Bill. He was a soft-spoken, sweet guy with a great sense of humor. He had a phenomenal memory for songs and poetry, played the harmonica, and sang beautifully. I have fond memories of after-dinner songfests at meetings overseas. We also sang at parties at his home in Boulder where he and his charming wife, Angela, welcomed us warmly.

But Bill never let his emotions show. Instead of blowing off steam at the directors' meetings at NCAR (as I frequently did), he would build up his frustrations internally until they exploded. He and I commiserated with each other regularly. He was under tremendous pressures from NSF RANN (Research for Applied National Needs), a new arm of NSF that was established for the transfer of technology to applications. Some three decades later it is difficult to understand the excitement that surrounded weather modification in the '60s and '70s. Several august advisory committees and reports of the National Academy of Sciences recommended major research on weather modification. In addition, a sizable number of commercial entrepreneurs conducted operational cloud seeding whenever there was a drought. But it was mainly the outrageous, but difficult to verify, claims of the Russians— that they were greatly reducing hail damage to crops in the Caucasus and elsewhere in the Soviet Union—that motivated the initiation of NHRE.

Because RANN looked upon NHRE as their golden opportunity to demonstrate an economically beneficial result, the RANN program manager was on the phone to Swinbank almost every day, trying to micromanage the project from Washington. This, and the implicit threats to cut the NHRE budget, put severe pressure on Bill. Indeed, he had just returned from a most disturbing meeting with the RANN program manager a few days before his heart attack and death during Christmas of 1973. It was only natural for many of us to suspect a relationship to his death. In view of the intolerable political pressures upon NHRE and my own natural skepticism about weather modification in general, it was probably foolhardy for me to accept the appeal from John Firor to assume the directorship.

NHRE was designed to be a combination research project to understand hailstorms and a quasi-operational hail suppression program. The fundamental hypothesis for suppression was that of "beneficial competition." In other words, one could seed the supercooled cloud portions of the storms to create more ice crystals and hail embryos than existed naturally so that the competition for water among the abundant embryos would prevent their substantial growth and thus permit them to melt on the way down. A second concept was to seed massively so that the silver iodide (AgI) nuclei would convert the supercooled cloud water to ice crystals. Upon collision with a hailstone embryo or hailstone, the ice crystals would bounce off, and so impede hailstone growth. But this concept was thought to be impractical because of the large quantities of AgI required to produce complete cloud glaciation.

In the Soviet Union, seeding was done by artillery targeted at the so-called "large drop" zone, thought to be comprised of large supercooled drops that could be readily nucleated and frozen. However, in the high plains of Colorado, where cloud bases are cold, the evidence indicated that some 90 percent of the hail embryos were snow pellets (graupel) grown by the riming of ice crystal cores by supercooled cloud droplets. Only 10 percent of the embryos were water drops (90, Sect. 7.3). Whether this was true in the Soviet Union was unclear. However, there was never any direct evidence of the existence of a "large drop" zone, since radar observations of high reflectivity aloft are ambiguous. In any case, artillery could not be used in the United States because of potential harm to commercial aircraft. Instead, we relied upon rockets fired vertically from aircraft flying below cloud base.

Using crop hail insurance statistics to deduce the frequency of hail in the four-corner area of northeastern Colorado (Wyoming, Nebraska, and Kansas), it was estimated that we would have enough storm samples for a statistically significant conclusion in five years if the average suppression effect were 20 percent. The latter number was based mainly upon crop hail insurance records in northeastern Colorado (91). However, crop damage is not directly proportional to the total mass of hail that falls. Damage to leaves is produced mainly by intermediate-sized stones, and therefore the amount of damage depends critically on the size distribution of hailstones. Hence, it is possible that a decrease in size of the large stones (if it worked at all) would produce greater crop damage.

When I assumed the leadership in early 1974 after three seasons of seeding, 33 storms comprised the sample, roughly half of them seeded and the others left unseeded as a control. I believe it was in the spring of 1974 that I went to Washington to report on the status of the program. I pointed out that there was less hail in the seeded group than in the unseeded ones, but that the difference was due to only one storm; thus, the results were statistically insignificant. However, a day or two later the RANN director reported positive results in testimony to Congress. We at NCAR were outraged, but our hands were tied.

We conducted the 1974 summer research and seeding program as before, but with some improvements in radar and surface observation techniques. After four months in the field, we returned to Boulder exhausted. We had precious little time to analyze the data during the next few months before getting ready for the next year's field program. Accordingly, I recommended that we stand down in 1975 and concentrate on data analysis as well as a reexamination of the basic hypotheses and seeding methods. I also recruited Keith Browning, then one of the gurus of severe storms, as chief scientist. He came from England to join us for the year starting in September 1974.

At that time, Colorado International Inc. (under Larry Davis) was conducting a hail suppression project at Nelspruit, South Africa. Their goal was to reduce the damage to tobacco crops, and their measure of effectiveness was in fact the observed damage in a sample of tobacco leaves as estimated by skilled tobacco appraisers. (Note that large leaf crops such as tobacco respond differently to hailstones than other kinds of crops.) Seeding was conducted by dropping silver iodide (AgI) canisters from a Learjet dive-bombing the cloud tops. They too had reported success. So I retained Prof. Joanne Simpson, a distinguished cloud physicist and enthusiast for weather modification from the University of Virginia, to accompany me on an inspection trip to South Africa in February 1975. We spent two weeks at Nelspruit, including an acrobatic flight in one of the cloud seeding missions, during which I came close to barfing. While I was turning green with nausea, Joanne was crawling around taking pictures from the windows. We came away quite impressed with the methodology, although we had no way of judging the statistical significance of the results. Subsequently, their claims of success were disproved by scientists from the Commonwealth Scientific and Industrial Research Organization (CSIRO) in South Africa.

In order to get a better picture of the status of hail suppression worldwide, Keith Browning and I organized an international conference on the subject that was held at Estes Park, Colorado in September 1975. Except for the Soviets, the meeting was well attended. A monograph on the conference proceedings was subsequently published (92). But the work that had the greatest impact was that by Browning and Foote (68), in which they presented a comprehensive analysis of a supercell storm. One of their major conclusions was that much, if not most, of the hail growth occurs above the "weak echo vault" where the strong updrafts provide an abundant supply of supercooled water to the hail embryos and stones that are near suspension for extended durations. Thus, the creation of more embryos by seeding would likely lead to increases rather than decreases in hail. Furthermore, Marwitz (93) had reported that 50 to 70 percent of the hail amounts occur on only 8 to 10 percent of the hail days. He also noted that just one storm day accounted for 70 percent of the total hailfall in the 1972 season in the Krasnadar Anti-Hail Project in the northern Caucasus. This large fraction was due to a major storm in which suppression attempts were unsuccessful. The implications were clear: Most hail damage was due to the large supercell storms that were either immune to suppression or would produce more hail as a result of seeding. The conference convinced many of us that we simply did not know enough about the natural hail growth processes to proceed with any degree of confidence. Also, the statisticians suggested that the detection of an average hail suppression effect of 25 percent with adequate confidence would have required at least several decades in northeastern Colorado (94).

It was about this time that RANN appointed a committee headed by a retired industrialist (Bernard Haber) to review NHRE. Because Haber was a management expert, the report of his committee emphasized inadequate management. Nevertheless, they expressed the hope that the strong leadership of David Atlas, who was still relatively new to the program, would improve performance. Evidently, this evaluation was unacceptable to the RANN program manager, so he appointed another review panel that would come up with conclusions more to his liking; and of course they did.

Among other things, one of their key recommendations was that NHRE should be divided into two components, research, and seeding, the latter to be done by some commercial firm, already wired to be done by Colorado International Incorporated, and using methods such as those used in South Africa. One of their arguments was that Atlas was not sufficiently committed to cloud seeding and was not a "believer." Indeed, I had already admitted that I was a skeptic. Although I would pursue the program vigorously to test the efficacy of hail suppression, I would have to be convinced by scientifically persuasive results. At that time, UCAR president Francis Bretherton and I were in nearly daily debate about what should be done. I had just about convinced him that we should stop the seeding program and return to basics.

But the debate did not stop there. UCAR convened an advisory panel of its own, made up of most of the distinguished cloud physicists in the United States and Canada. They included Lou Battan (Arizona), Roscoe Braham (Chicago), Roland List (Toronto), Peter Hobbs (Washington), and Joanne Simpson (Virginia). Bob Fleagle (Washington) represented the board of trustees. The meeting was held on November 12, 1975. (This was the day after I had returned from San Francisco, where I had said farewell to the Chinese scientific delegation with whom I had spent most of the prior three weeks in exhausting travel around the country as president of the AMS and host to the delegation.) One by one, the panel members credited me as a scientist and leader who had the courage of his convictions. But with a couple of exceptions, they went on to emphasize that NCAR must proceed with the seeding program since it was the largest and most visible project of its kind in the world, and that weather modification would be set back for years if the program were stopped then. These were not scientific arguments but political ones. Nevertheless, they prevailed in a closed meeting of the panel that night from which I was excluded. It was clear that I could not stay on as director under such circumstances, so I resigned two weeks later.

In my farewell talk to the NHRE staff, I asked them to continue their best efforts on the program in behalf of NCAR. But my emotions got the best of me and tears flowed.

I spent much of 1976 writing a comprehensive review of hail suppression in particular and weather modification in general (95). My findings may be summarized as follows: (1) the reports of both negative and positive effects of seeding suggest that there are a variety of conditions in which seeding produces increased hail; (2) such variability is evidently independent of the seeding method (which some investigators had suggested was crucial); (3) there are at least four physical mechanisms that may produce increased hail; (4) the results achieved under one set of conditions or in one part of the world are not transferable to other conditions or other places; and (5) the suppression of hail may be negated by a more economically valuable loss of rain. I do not know whether these arguments (most of which were available prior to the 1977 publication) were decisive, but these and others finally held sway in early 1976 when NCAR decided to abandon the seeding program. The latter decision was a bittersweet vindication when, by mid-1976, I had decided to accept the invitation of NASA to assume the directorship of what became the Goddard Laboratory for

Atmospheric Sciences at Goddard Space Flight Center. A number of the advocates of weather modification have since held me responsible for the decline of research in this field. If that is the case, then I am flattered, for it evidently resulted in a significant rethinking of the subject. In rereading the 1977 "Paradox" paper today I was pleased to find that most of it is still relevant.

In view of the trauma surrounding NHRE and the sadness with which I left my friends and colleagues, I was greatly comforted by a letter to me from Edwin Wolff, assistant to John Firor, immediately after my resignation. I retain this among my most cherished mementos.

After my resignation from NHRE, I joined the advanced study program and moved my office up to the "isolation" tower with distinguished colleagues such as Chuck Leith and AMS past presidents Bernard Haurwitz, Phil Thompson and Will Kellogg. This was an exclusive little club. Our kaffeeklatsches included wide-ranging discussions of meteorology, the history of science, and politics. And it was enjoyable to listen to the recorder concerts by Will Kellogg and his friends during lunch.

In addition to my reprise paper on hail suppression, Henry Gertzman joined me to do an interesting paper on sampling errors in the measurement of rain and hail (96), while Carl Ulbrich and I completed another on the use of path-integrated microwave attenuation and differential attenuation at two polarizations to measure path average rainfall (97). We also demonstrated the near-linearity between the attenuation and rainfall rate at wavelengths of 0.9 to 1.8 cm. The conclusion of the latter paper includes the simple sentence: "The method may be extended to paths between the ground and aircraft or satellites." This presaged the adoption of the surface reference technique (SRT) for measurements of path integrated rainfall or attenuation (PIA) by the precipitation radar on the Tropical Rain Measuring Mission (98, 99).

However, the most pleasant undertaking was the paper I did with Charles Elachi and Walter Brown at the Jet Propulsion Laboratory at California Institute of Technology. Elachi had invited me there to discuss a proposal to use a very simple radar that was to be dropped into the atmosphere of Jupiter from the Galileo satellite to determine if there was any precipitation in its atmosphere. I was astonished and amused that the mission design limited the radar to a total of about 5 watts and only 3000 bits of data in a 30 sec traverse of the Jovian atmosphere. The proposal we prepared was not accepted.

During our discussions, however, Elachi showed me an image taken with a 3 cm wavelength synthetic aperture radar (SAR) on board the NASA Convair 990 over the Pacific Ocean off the coast of southeast Alaska. The wide vertical SAR beam looks out mainly to one side of the aircraft, so it was odd to find that the image showed patterns similar to those seen on vertical radar cross sections through precipitation. Although we continued our discussions of the Galileo mission through lunch, I could not get the SAR picture out of my mind. And then the light dawned. Since the SAR processing system is designed to look at the ground and the ocean, it must filter out all but those echoes that have near zero radial motion or Doppler velocity relative to the airplane. Thus, the radar will see falling precipitation on the upwind side of the aircraft where the approaching horizontal component of the precipitation just cancels out the receding component due to the fall-speed. The net result is essentially a near vertical cross section through the precipitation that is displaced to the upwind side of the aircraft (100). This was one of those exquisite eureka moments of discovery that make science a joy. And it further confirmed my view that the scientist must get away from home to broaden his perspective and refuel his intellectual engine.

CHAPTER 7

American Meteorological Society

My election to president-elect of the AMS in 1974 and president in 1975 was a mixed blessing because of the traumatic year I was to experience as NHRE director in 1975. I had previously served on the Council of the AMS in 1962–1964 and 1972–1974 and found it an enjoyable and rewarding experience. In the earlier period, it was a special pleasure to meet and work with some of the old-timers in the field such as Pettersson, Neiberger, Tom Malone, "Shorty" Orville (father of Harry and Dick Orville), Verner Suomi, Lou Battan, John Beckman, and many others. Later, it was leaders like Pat McTaggart-Cowan from Canada, Dave Johnson, Joe Smagorinsky, Dick Reed, Will Kellogg, Werner Baum, and Earl Droessler.

My all-time favorite is Ken Spengler, the perennial executive director, now emeritus, of the Society. And I must not forget Evelyn Mazur, Ken's top assistant. The lion's share of the work was done by Ken and Evelyn and a mere handful of staff through the mid-sixties. Ken is a gentle, thoughtful, and considerate man who knew almost everyone in the Society in the early days. Everyone loved him then, and those who know him now still do. He is a true statesman who treated all members, no matter where they stood in the professional hierarchy, with dignity and warmth. And Evelyn was the prototype of efficiency and good humor. Indeed, through the 1960s, the AMS was a fraternity and professional society in equal parts. Many of us had met during the war either as students, teachers, or military officers. Our meetings were of reasonable size so that it was easy to get to know one another, and many of us were on a first-name basis. Attending a conference was truly a reunion.

The highlight of 1974 was the visit of the AMS delegation to mainland China. After President Nixon opened relations with China in 1972, Dick Reed, then AMS president, had contacted his Chinese counterparts in the interest of restarting relations between our two societies and making way for more substantive scientific exchanges and cooperation. It took almost two years to get a response; this led to the invitation to visit them. The delegation, one of the first non-government groups to visit the People's Republic of China (PRC), was comprised of then president David Johnson, our delegation leader; past presidents Will Kellogg and Dick Reed; president-elect Dave Atlas; Ken Spengler; and our wives (with the exception of Dave Johnson, who was then a widower). (Plate 22) We were assigned a caravan of five cars, one for each couple, and one for Johnson. In 1974, in the midst of the Cultural Revolution, when bicycles were the normal mode of transport, a line of cars meant distinguished visitors, so that we were gazed upon with awe and wonder wherever we drove. When we sometimes eluded our guides, it was fun to walk through the streets, where we exchanged greetings with the locals.

We visited the Institute of Atmospheric Science (Academia Sinica) and the Central Meteorological Bureau. We also visited schools, factories, communes, and a hospital; attended the theater; and were feasted sumptuously. One of the main benefits was getting to know the leaders of Chinese meteorology. On the government side, we met Zheng Ming Tsou (VP of the Chinese Meteorology Society), our main host (since deceased). On the scientific side we renewed acquaintance with Du Tzeng Yeh and Yi-Ping Hsieh, both of whom had gotten their doctorates at the University of Chicago in the late 1940s.

Our trip has been described thoroughly in the *Bulletin* of the AMS (BAMS) (101). We had serious disagreements as to how to handle the story because we did not wish to insult the Chinese by describing some of the primitive aspects of their science and technology. Nor did we wish to recount the rampant and repetitive singsong propaganda that was presented wherever we went. Thus, the account published in the *Bulletin* did not reflect a realistic picture of what we saw in the PRC in 1974.

Before we left the United States in mid-April, some 140 tornadoes broke out across the midwest section of the country on April 3–4, 1974. This was a disaster of unprecedented proportions. The AMS Executive Committee prepared a powerful statement to Congress and the state governors recommending prompt action on the development of Doppler radar and severe storm warning methods. We left that statement with one of the members of the Executive Committee who was also a senior civil servant in the National Weather Service. As a notoriously cautious government employee, he made the mistake of submitting it to the NOAA legal counsel. Although the statement was to be released over the imprimatur of the AMS, our colleague was told that he would be subject to a fine and/or imprisonment if he distributed the document. Thus, when we spoke to him from our assembly point in Hong Kong, we found that the statement had not and could not be distributed. This illustrates one of the constraints under which the AMS operated. We often could not take action on behalf of the public because so many of our members and leaders were government employees forbidden from lobbying.

Because the AMS presidency is limited to one year, a president must choose priorities carefully. I chose to focus my term on the issue of atmospheric science and public policy. This was a period in which all of basic research was being questioned. The general response from the research community is typified by the statement of Harvard Prof. Harvey Brooks that basic science will have to be justified on the grounds that "science is a seamless web, such that the 'useful' parts cannot prosper unless the apparently 'useless' parts are well supported." In a BAMS editorial, I had written that "basic research may be regarded as insurance: 'accident' insurance in the event that the particular research approach we have chosen in pursuit of an early [applied] solution should fail, and an 'endowment policy' to assure that our scientific offspring have the tools with which to solve the problems of the future" (102). This was followed by several other editorials and rebuttals on the subject, particularly related to the issue of overselling potential applications and thereby endangering our credibility. Those arguments are still pertinent today.

In order to press forward in the public policy arena, we invited the congressional subcommittee on the environment and the atmosphere to attend a special session of the 1975 Conference on Severe Local Storms in Norman, Oklahoma. Congressmen George Brown and Lawrence Winn, and members of their staff attended. This was the first time that members of Congress had attended a meeting of the AMS. The speakers included professors Ted Fujita (Chicago), Lou Battan (Arizona), Allen Pearson (NWS), Rick Anthes (Penn State), and myself. The subject matter included virtually all of the needs of the time: the near-term promise of tornado detection and warning systems such as Doppler radar, the importance of higher-density surface and radiosonde observations, and the longer-term expectations from mesoscale modeling and research. It was an exciting session and our congressional guests left with increased understanding of the problems, the potential for short-term solutions, and the advances expected from longer-term research.

We followed up this initiative with a symposium on atmospheric science and public policy at the January 1976 annual meeting of the AMS in Philadelphia (103). At that meeting, we had some of the key thinkers on science and public policy in the country. They included Congressman George Brown on "Environmental Quality and the Role of the Congress," Dr. Lewis Branscomb (then chief scientist of IBM) on "The Role of Institutions in Rational Policy Making," Dr. Walter Orr Roberts on "Science, Technology, and the Changing Human Condition," Dr. William Carey (then executive officer of the American Association for the Advancement of Science) on "Science in a Negotiating Society", Dr. Robert White, NOAA administrator, on "Deciding Public Meteorological Policy," and Prof. Harvey Brooks, professor of technology and public policy of Harvard University, on "Atmospheric Science and Atmospheric Politics." The talks were followed by a panel discussion that included the speakers and Tom Malone, Lou Battan, Bob Fleagle, and Seville Chapman. With minor changes of the scientific issues, the record of the presentations is still relevant today. In 1976, they included such problems as the possible effect of supersonic aircraft on stratospheric ozone and the subsequent impact on skin cancer and agriculture, severe storm detection, the climate problem (cooling or warming), and the nuclear option in lieu of fossil fuels that produce CO_2. But the dominant enigmas were how to make policy decisions on the basis of insufficient scientific evidence, and how to separate value judgments from scientific issues.

We had also arranged that our 1976 annual meeting would be a reunion of past presidents where they would be presented with a memento comprised of a barometer, a thermometer, and a plaque indicating the year of their presidency. It was a nostalgic meeting attended by 12 past presidents and several future ones. Plate 9 is an historic photo of the 12.

Although we had merely started the public policy process, the Council and I were pleased with the steps we had taken. My activity as AMS president provided me with a sense of satisfaction that made up in large part for the NHRE ordeal. On balance, I emerged from 1975 with an optimistic outlook.

Traditionally, the AMS president is also a delegate to the quadrennial Congress of the World Meteorological Organization. So, I was a member of the 1975 delegation that was headed by Bob White, NOAA administrator at the time. These meetings were a mixture of operational, scientific, and political issues, with much behind-the-scenes negotiating about the choice of the next director general. Many national delegates came prepared with position papers on the issues; these were debated behind the scenes among interested delegates who revised them repeatedly to modify all the ifs, ands, and buts, and then printed them in five languages for distribution. I spent several days revising the position paper on weather modification with input from a couple of other nations. During the second week Bob White asked me to present the U.S. position on weather modification. But the delegates from Australia, the United Kingdom (Dr. John Mason), and several other nations were recognized before me. When my turn came, much of what I had to say had been preempted by prior speakers. So I began by stating: "Much of what I had originally planned to say has been preempted by the distinguished delegates from Australia and the United Kingdom. In fact, the remarks of Dr. John Mason were so similar to my own that I suspect he had access to my notes," and proceeded with a shortened version of the prepared paper. Well, I thought Bob White would have apoplexy. But at a social gathering that evening John Mason joked that he was going to sue me for several hundred million dollars for having insulted him in five languages simultaneously (the official ones of the WMO).

As a sidebar, I was concerned that the AMS anticipate the need to train a replacement for the "aging" Ken Spengler, who was 59 years old in 1975. I discussed this idea in a confidential meeting of the Executive Committee in Chicago and swore the committee members to secrecy. Within 15 minutes of the end of that meeting, Ken had learned the secret and kidded me about my trying to fire him. As I write this in 2001, at age 77, I yearn for my 59-year-old youth, and I marvel at Ken's good health and vigor 25 years later. He never lets me forget it.

The American Meteorological Society has been good to me. At the 81st Annual Meeting in Albuquerque, I was elected an honorary member. It is awesome to be listed among those whom I have revered through the last half century.

CHAPTER 8

NASA Goddard Space Flight Center Goddard Laboratory for Atmospheric Sciences, 1977–1984

During the summer of 1976, while I was recuperating from the NHRE fiasco, I had two visitors from NASA: first Bill Nordberg, director of the Applications Directorate of Goddard Space Flight Center (GSFC), and then Bob Cooper, the Goddard director. Bill was there to recruit me to head the severe storm program at GSFC. Both were dynamic personalities, friendly and enthusiastic. Nordberg made a fine pitch; both NASA and the job looked very attractive. I was later overwhelmed to find that he was suffering from an inoperable brain tumor to which he succumbed just a few months later. A couple of weeks after Nordberg's visit Cooper came by to make an even more enticing offer—to establish a broad-based atmospheric and oceanic science laboratory. He also made the unprecedented commitment to permit us to hire without limit—a promise that he made on the bold assumption that he could sell the idea to NASA headquarters. To my amazement, he did. At the time, I was also entertaining an offer to head the Atmospheric Science Division at NSF. After considerable thought and confidential discussions with my colleagues, I accepted the NASA offer and arrived at Greenbelt, Maryland in January 1977, simultaneous with the start of the Carter administration. Lucille was anxious for the move back to the East Coast because she wanted to be close to her mother. Our move was also a stroke of good fortune because Lucille had a malignancy that had been misdiagnosed in Boulder but was surgically cured in Maryland. Here is a life-and-death case of serendipity.

In early 1977, the meteorological activity at Goddard comprised some 60 people in the Atmospheric and Hydrospheric Applications Division (AHAD) under the direction of William Bandeen. The division was also supported by some 100 on-site contractor staff. AHAD covered an exceedingly broad spectrum of research and satellite projects. It included modest-sized groups dealing with hydrology and oceanogra-

87

phy, stratospheric chemistry and solar terrestrial physics, severe storms and large-scale weather systems. The division also enjoyed the support of the Information Extraction Division for software and data processing, the Earth Observations Systems Division for instrumentation, engineering, design, and fabrication, and the Communications and Navigation Division for microwave sensor design and development.

Staff members were project scientists on a variety of satellite instruments such as Nimbus 5 and 6 (John Theon), Nimbus G (Bill Bandeen), Tiros N (Albert Arking), synchronous meteorological satellite (William Shenk), the radiation budget (Bob Curran), and more. Goddard also supported a number of university investigators such as Verner Suomi at Wisconsin, Ted Fujita at Chicago, Tom Vonder Haar at Colorado State, Zdenek Sekera at UCLA, Ben Herman at Arizona, and others. These were important synergistic collaborations; however, the relationships were weakened when NASA headquarters assumed direct program management and the university investigators no longer had reason to report directly to their project counterparts at Goddard.

Until 1977, NASA operated under a "lead center" management style in which each of the centers had responsibility for a particular discipline. Goddard was the lead center for meteorology. Morris Tepper was the overall program manager at headquarters. His operational arms were the Meteorological Program Office at Goddard under Harry Press and AHAD under Bandeen. Thus, Goddard staff had remarkable power in directing the distribution of funds to the various NASA centers and to university investigators. Naturally, we were approached frequently by supplicants for funds. But that ended when the new administration pulled the management role into headquarters, dividing each discipline into subspecialties under the control of program managers. The success of this practice depended upon the style of the individual manager. Most relied upon the center scientists for grassroots planning, but a couple misinterpreted their functions and ruled autocratically from the top down. Call them micromanagers; I think of them as "abominable no-men" whose power resided in saying "no" rather than in finding the means of helping us get the job done.

Each specialty area was defined by a research and technology operating plan (RTOP) that outlined the plans of the various investigators. Full-blown proposals were not yet required. As lab chief, I had the flexibility to realign up to 20 percent of the funds within each

RTOP in order to attack exciting emerging ideas or to combine related activities in a symbiotic fashion. However, some at HQ thought that this relinquished too much control to the centers. Therefore, they reduced the size of each RTOP and divided them into smaller pieces so that we at Goddard would not have excessive flexibility to deviate from planned activities. This was unfortunate, because almost none of the headquarters managers had a broad view of the overall research program, nor were most of them up-to-date on the science. In fact, at Goddard, I was probably the only one who had a comprehensive view of the laboratory-wide program because I read and critiqued virtually all the RTOPs (later, proposals) submitted by our staff. An unfortunate result of the takeover of detailed funding control by HQ was that individual scientists would go directly to HQ to sell their project ideas, thus undercutting my role as laboratory director. What a way to run a ship! I'm afraid that the tug-of-war between NASA HQ and the center managers still continues. Incidentally, one of the favorable practices at HQ is to bring in "rotator" program managers who remain for a couple of years before returning to their home centers. In this way, they remain scientifically competent and credible.

Until the bugs in the new management scheme were smoothed out, conflict between Goddard and HQ managers was rampant. One result was the difficulty of recruiting a new branch head for the severe storms program. While it was common courtesy to advise the HQ staff of the candidates, the final decision for hiring rested with us at Goddard. It was only after the resolution of this conflict that I was able to attract Joanne Simpson away from her professorial chair at the University of Virginia to head the severe storms program in 1979. She became the intellectual center of gravity for that program, attracted other first-rate scientists, and led a magnificently imaginative and productive research agenda. She went all out in support of her staff and they responded with utmost loyalty. She also had the reputation and per-suasive power to attract necessary resources from HQ.

Another complication of NASA management was the advisory committees, comprised mainly of scientists from academe. They had tremendous influence because HQ managers paid homage to them. On the other hand, they often did not show corresponding respect for NASA scientists. I had a problem with this pattern of behavior because many of the committee members were my colleagues from universities or NCAR so that I did not hesitate to question their judgment and engage them in discussion. Moreover, the advisory committees frequently mistook their advisory role for management and undercut the independent actions of the NASA scientists who had the responsibility for planning and implementation. This kind of second-class citizenship for government staff remains a problem today. In his wonderful book on the history of NASA, Homer Newell (104) describes the love–hate relationship between university and NASA scientists and managers:

> The complacent assumption of the superiority of academic science, the presumption of a natural right to be supported in their researches, the instant readiness to criticize, and disdain which many if not most of the scientists accorded the government manager, particularly the scientist manager, were hard to stomach at times. . . . [The first NASA administrator, Keith Glennan] could not restrain an outburst at the arrogant presumptuousness of the [academic] scientists . . . [and] the view that NASA scientists are second-rate by and large.

While at NCAR I also dealt with advisory committees for each of the facilities in ATD. However, they understood that we were responsible for management and could not simply knee-jerk react to all their advice. The entire issue of institutional management of science and technology deserves much more serious attention than is possible here. I subscribe to the philosophy of Lewis Thomas (105) that is summarized in Chapter 12.

My goal in 1977 was to build upon the existing strengths of AHAD and to transform it into a world-class research laboratory. My colleagues quickly recognized that I was attempting to build an "NCAR East," which was not far from the truth. I worked hard for several months to prepare a program plan focused around the scientific activities with a strong statement of research philosophy and organization. The plan was approved, and the Goddard Laboratory for Atmospheric Sciences (GLAS) was established in October 1977.

In line with the concept of a major laboratory, it had already been planned to transfer the global weather modeling activity from the Goddard Institute of Space Sciences (GISS) in New York to Greenbelt. This group, which became the Global Simulation and Modeling Branch under Milton Halem, already included some top-notch people—such as Yale Mintz (one of my instructors 35 years before), Eugenia Kalnay, Jagadish Shukla, and Robert Atlas (no relation to me). They also had an ongoing agenda, which gave us new momentum from the very start. And they had excellent scientific staff support from a contractor.

We started an intensive recruiting campaign in the fall of 1977. Our appointments committee was comprised of my deputy, Bill Bandeen, the five branch heads, and myself. Our style was patterned partly after that of the University of Chicago and NCAR. We aimed at unanimity in the selection of new staff. It was interesting to note the difference in standards between those who had come from academe and those who had spent most of their careers in the government. The latter, having been deprived of new staff for years, were naturally less selective; the others insisted on a record of high achievement and strong letters of recommendation. By 1980, we had hired 35 new scientists, increasing the staff to a total of 96 not counting contractors. All but two had Ph.D.s, and these two had come with the highest recommendations.

Twenty years later, it is gratifying to note the remarkable success attained by the vast majority of those we hired. Joanne Simpson stands out; in addition to leading the Severe Storms Branch, after my retirement as lab director she became the associate director for meteorology of the Earth Science Directorate and the project scientist for the Tropical Rainfall Measuring Mission. She has been showered with a series of well-deserved high honors. Louis Uccellini distinguished himself as leader of the visible and infrared spin scan radiometer (VAS) sounder studies on the geostationary operational environmental satellites (GOES). He and his group later conducted a seminal synoptic analysis of the President's Day snowstorm. It was one of the first to use mesoscale numerical model output as a basis for the synoptic study. This work culminated in an AMS monograph on winter storms. Uccellini moved to NOAA as director of the office of meteorology and then became the director of the National Centers for Environmental Prediction (NCEP).

Eugenia Kalnay followed Milton Halem as head of the Global Modeling and Simulation Branch, then moved to NCEP as director of research, and is now chair of the Department of Meteorology at the University of Maryland. Jagadish Shukla moved to the University of Maryland and then formed the Center for Oceans, Land, and Atmosphere (COLA), of which he is president. Gerald North received a distinguished professorship at Texas A&M University, where he continues an outstanding career in climate studies; several others are also in academe. After a stint at NCEP, Paul Kocin is a winter storm expert at the Weather Channel.

Others who remained at Goddard have also gained renown. Claire Parkinson has achieved recognition as a science author and an expert on sea ice, and now as project scientist of the Aqua satellite of the Earth Observing System. After having gained distinction in remote sensing of aerosols and their effect on climate, Yoram Kaufman did a top-notch job as project scientist for the Terra satellite launched in 1999. And Michael King has become the senior scientist and leader of the Earth Observing System after an extraordinary career in radiation and climate. Milton Halem is now associate director of information systems at Goddard, after an extended career as director of the super-computer center. Bill Lau, now head of the Climate and Radiation Branch, has achieved recognition in a broad range of climatic problems. And Anthony Busallachi also reached great scientific heights as a modeler of the oceans and their interaction with the atmosphere, having served as the chief of the Laboratory for Hydrospheric Sciences until his recent departure for a professorship at the University of Maryland. In short, our appointment process was on target; there were remarkably few duds in the pack. Others who were in AHAD prior to the formation of GLAS have also gone on to major achievements. These include Tom Wilheit, an expert in microwave remote sensing, who is now a professor at Texas A&M, and Robert Meneghini, who has been one of the prime movers in bringing the TRMM satellite to reality.

The fact that so many of our top scientists moved on to more prestigious positions at other institutions is the result of natural evolution. It reflects well on the stimulating and nourishing environment that we were able to establish within the Goddard Laboratory for Atmospheric Sciences, and within Goddard and NASA in general. We were fortunate to have some highly enlightened directors during my tenure and in the years that followed (Tom Young, Noel Hinners, and John Townsend). And, with few exceptions, we benefited from supportive program managers at NASA headquarters. Unfortunately, the halcyon days of liberal hiring did not last long, and it has not been possible to infuse enough new blood into the institution to replace those who departed for greener pastures and higher positions. Indeed, it is surprising that Goddard has maintained its good reputation in spite of the tight constraints on hiring and promotion.

The glory days of liberal hiring by GLAS came to an end with the 1980 election of President Reagan, who froze all hiring the day after his election. We thus lost 15 scientists to whom we had made offers of positions. Among these was Tzvi Gal-Chen, who had joined us as a contract scientist and was about to become a U.S. citizen. Tzvi went on to a highly productive career at the University of Oklahoma. It was a great loss to the community when he died at a relatively young age.

My role as a laboratory chief was greatly aided by the remarkably versatile, dedicated, and tactful associate chief, Bill Bandeen. He was efficiency personified. With his long tenure at NASA, he knew his way around and taught me the ropes. He assumed the responsibilities for the myriad activities for planning, budgeting, and reporting. He also shared one of the key functions: holding the hands of those staff who felt they deserved promotions or had other gripes. But I made it a point to present the GLAS candidates for promotion at the periodic meetings of the directorate where we were in competition with other laboratories for promotion slots. Bandeen's collaboration also made it possible for me to focus my attention on the core scientific areas both at Goddard and at HQ.

This was a very exciting period for the space program. While the preceding years were built largely on the beautiful pictures of clouds and storms and their qualitative use in forecasting and warning, the late 1970s started to bear fruit in the quantitative measurement of atmospheric parameters. Prof. Ted Fujita at the University of Chicago had provided major insights into the structure and evolution of severe storms, including hurricanes and tornadoes, based upon cloud patterns and motions. And Prof. Verner Suomi and his staff at the University of Wisconsin had developed the Man–Computer Interactive Data Analysis System (McIDAS) that permitted the rapid analysis and manipulation of satellite images. Subsequently, we developed an improved version called the Atmospheric and Oceanic Information Processing System (AOIPS), which continues to serve our needs admirably today.

The 1978–79 First GARP Global Experiment (FGGE) program had been motivated in large part by the new ability to obtain soundings of temperature and humidity by means of the high resolution radiation sounder (HIRS) (first flown on Nimbus 6), the microwave sounding unit (MSU) flown on the TIROS satellites, and by the use of GOES satellites to obtain winds from cloud motions. The expectation was that such information, obtained on a global basis rather than by radiosondes based over the populated continents, would permit the skillful prediction of weather to extended periods. Accordingly, much of the work of the Global Modeling and Simualtion Branch was focused upon assessing the impact of the new data on forecast skill. As it turned out, most of the increased skill was found to occur in the Southern Hemisphere, where radiosonde observations were sparse. Nevertheless, these studies yielded a host of significant insights about extended weather forecasting. Shukla collaborated with Moura from Brazil in the use of one of the early climate models to study the effects of shifts in ocean surface temperatures in the Atlantic Ocean on droughts in northeastern Brazil. And Mintz and Shukla explored the influence of land surface evapotranspiration on climate.

Meanwhile, Uccellini's mesometeorological team was conducting experiments with the VAS to determine its utility in initiating mesoscale prediction models of severe storms. Fritz Hasler had started his fascinating studies of the applicability of stereography from GOES to the detection and warning of severe storms. Shmugge was doing seminal work on the measurement of soil moisture by microwave radiometry.

Walsh and his colleagues also had a superb airborne surface contour radar operating at 35 GHz with which he could measure directional wave spectra over the sea. Zwally and colleagues (106) had conducted a masterful analysis of the annual cycle of sea ice in both the Antarctic and Arctic during which they discovered the Weddell Polynya, a huge (200,000 km^2) open water area in the midst of winter. Claire Parkinson subsequently explained the polynya through the use of a numerical model (107). Robert Bindschadler, a young glaciologist, was providing new insights into the movement and calving of glaciers. He has recently been elected president of the International Glaciological Society. And Norden Huang and colleagues did pioneering analyses of ocean topography with satellite altimeters as well as studies of the ocean wave spectral shape and the statistical properties of the ocean surface. Norden has recently been elected to the National Academy of Engineering. This is but a taste of the stimulating environment that characterized GLAS in the early 1980s.

It was a great learning experience for me to hear about these activities at the weekly seminars that I attended regularly. Although I was a neophyte in many of the subject areas, I was not reluctant to ask questions. Sometimes I did this out of simple curiosity; at others I did so to trigger discussion. As is often the case, one question led to another and the discussions frequently became vigorous. Our seminars were among our most enjoyable activities.

In line with my emphasis on interinstitutional dialogue, I got a great kick out of exploiting an unused communication satellite parked in space to hold a number of joint seminars between NCAR and GLAS. Not only were the discussions effective, it was also fun for me (the moderator) at Greenbelt to call upon members of the audience in Boulder to ask them questions or seek comments. Considering the ease with which the television industry holds live discussions around the globe, it is disappointing that the scientific community has not yet taken advantage of this technology for nationwide and intercontinental conferencing.

What I learned from the seminars, reading, and from extended discussions with the staff scientists sufficed to provide me with the knowledge to represent the laboratory's work to upper management and at scientific meetings. It also triggered my own ideas so that I was able to join the scientists and engineers as a coauthor. Thus I collaborated with others on a variety of space-related papers on topics such as: (1) a futuristic operational meteorological satellite system to provide weather and climate information; (2) a method of estimating heat transport over coastal waters during cold-air outbreaks; (3) the use of geosychronous satellite IR observations to estimate thunderstorm rainfall; (4) various applications of space-borne lidar for weather and climate; and (5) broad overviews of remote sensing for weather and climate. Goddard director Bob Cooper also asked me to lead a climate study team that produced a major report, "Climate Observing System Study," or COSS.

In addition, in cooperation with NOAA National Environmental Satellite Service (NESS) and the National Climate Program Office, we held a workshop on precipitation measurements from space in 1981. It is interesting to read the summary of that meeting now because most of the recommendations have been implemented under the Tropical Rainfall Measuring Mission (TRMM) (108). It was at the latter workshop that we presented the study on the prospects for precipitation measurements from space (109). This was followed by a series of papers on multiparameter measurements of rainfall (110), and the theory and tests of the surface reference technique (SRT). The latter method has become the basis for rainfall measurements by the radar on board TRMM. The team of investigators on these studies included Bob Meneghini, Carl Ulbrich, Kenji Nakamura from Japan, and myself. (Plate 23)

An interesting sidebar to the above is the story of how we got involved in a joint project with the Japanese to develop TRMM. From my contacts in Japan, I had learned that the Communications Research Laboratory (CRL) there had developed an airborne dual-wavelength radar/radiometer with matched beams operating at 10 and 35 GHz. Since such a system was close to that which we needed to implement the SRT method, I contacted Dr. Keikichi Naito (who had worked with me many years before at AFCRL) and asked him to pass my letter to Dr. Nobuyoshi Fugono at CRL. I chose this circuitous approach because official correspondence had to go through the NASA Office of International Affairs, and I feared that they would either block or delay a formal letter. The letter suggested that we initiate collaborative experiments by installing their system on a NASA aircraft and that they also send a scientist/engineer to accompany the equipment. I pursued the project in discussions with Kenji Nakamura at the 1983 Conference on Radar Meteorology in Edmonton, Alberta. The idea was received favorably in Japan, and they initiated the formal correspondence with NASA to suggest the collaboration. Since this generous proposal came from Japan, the program manager at HQ could hardly say no. Nakamura and the equipment arrived in 1984, and he remained for two years. From then to the present day we have had a series of visiting Japanese scientists with whom we have developed close friendships and productive collaborations. When North, Wilheit, and Thiele formulated the TRMM proposal shortly after my official retirement, it was only natural that the Japanese would want to join us. Indeed, it would not have been economically feasible otherwise, for the Japanese assumed the responsibility for the development of the radar and the launch of the satellite in November 1997.

Notable byproducts of the U.S.–Japan collaboration were the book *Spaceborne Weather Radar* by Bob Meneghini and Toshiaki Kozu, and the doctorate awarded to Meneghini by Kyoto University.

Bob Meneghini and I joined in a variety of subsequent papers dealing with the feasibility of spaceborne rainfall measurements, both theoretically and experimentally. Although I may have started out as his mentor, the tables quickly turned so that he became mine. He has often taken a simple scheme of mine, extended and examined it rigorously, and simulated its performance mathematically. He led the development of the algorithms for the retrieval of rain from the TRMM precipitation radar with collaboration from a number of our Japanese colleagues. And he has successfully processed the space data to produce remarkable rainfall measurements over the tropics. Indeed, the spaceborne radar is so stable that it has also become a prime tool for the calibration of ground-based radars around the world. Bob Meneghini is quiet and modest, but there is no doubt that he stands near the top in the radar meteorology hall of fame.

Two other projects that we undertook in the last two years before my official retirement from NASA are worth mentioning. I was fascinated by the cloud streets that developed off the East Coast during cold-air outbreaks; these are prominent in satellite imagery during the winter. The upwind boundary of the family of cloud streets was an image of the coastline. Surely, this pattern must contain some meteorological information. We did a modest study that showed that the distance between the coast and the upwind boundary of the clouds was related to the heat flux from the ocean to the atmosphere; the smaller the distance, the greater the flux. Since we had an airborne lidar capable of mapping the structure of the boundary layer and airborne radars capable of measuring directional ocean wave spectra, it was natural to attempt a coordinated experiment involving all these remote sensors. Thus, we conducted a major field program—the Mesoscale Air–Sea Exchange (MASEX) project—during the week of 16 January 1983 off the Atlantic coast between Virginia and New York. All that we required in addition to the Goddard resources was an aircraft with the full complement of in situ meteorological sensors and dropsondes. This we got through informal negotiations with our friend Frank Marks at the Hurricane Research Division (HRD) of the NOAA Atlantic Oceanographic and Meteorological Laboratory; fortuitously, they had not exhausted all the aircraft flying time allocated to them. I also invited Dr. Dag Gjessing, head of the Norwegian Remote Sensing Technology Program, to send a scientist with his multifrequency continuous wave radar for installation on one of the three aircraft. All this was arranged within a couple of months without the knowledge of NASA HQ. Luckily, we had four strong cold-air outbreaks during the week of our campaign.

The results of this experiment were thrilling (111). In essence, the lidar revealed the structure of the convective roll vortices within the unstable marine boundary and across the inversion. We were able to match that structure to the fine-scale wind, temperature, and humidity measurements made by the NOAA P-3 aircraft. Among other things, we found that the rising arm of the roll vortex coincides with a column of high lidar reflectivity and with the updrafts that transport aerosols, moisture, and heat up from the surface. The roll vortices also serve to produce low-level convergence, which concentrates the small-scale buoyant eddies near the surface to form a single well-ordered updraft. Moreover, contrary to prior knowledge, we found that the downward flux across the inversion was due mainly to the entrainment of warm eddies that are then transported downward by the large-scale roll vortex circulation. Our ocean radar colleagues also found beautifully organized directional wave spectra developing downwind in excellent agreement with theory. This project was as close to a laboratory experiment as one could possibly conduct in the atmosphere. Although the experiment was conducted while I was still the GLAS director, the analysis and publications were completed under an NSF grant while I was employed at the University of Maryland.

The details of this experiment are not so important here. Rather, I wish to emphasize the importance of institutional flexibility in permitting scientists to follow their own instincts and to exploit unique opportunities for interdisciplinary and interinstitutional research. We were able to do all this because we had most of the tools within our own organization and had friendly collaborators in other institutions who also exploited their own flexibility. MASEX would have required endless, frustrating planning and negotiation had we gone through formal channels. Also, this kind of work speaks to the advisability of having a hands-on scientist as a laboratory director, one who can gather the scientists from various disciplines within his own organization—who may not even be aware of one another's work—in a coordinated attack on an interdisciplinary problem.

One of the last studies that I undertook before retiring as laboratory director was again the result of serendipity. I was visiting NCAR in 1983 when Peter Hildebrand took me by the collar to show me some of the first observations of the motions of precipitation taken with the airborne ELDORA Doppler radar over the ocean off the coast of Washington. He was speaking with the kind of enthusiastic excitement that has marked my own reaction to novel observations. But while he was describing the wind field *above* the ocean surface, I was entranced by the interesting color-coded Doppler velocity patterns *below* the ocean surface. It didn't take long to recognize that these were mirror images of the motions as seen by reflection from the ocean surface to the precipitation and back.

Tom Matejka and I (112) subsequently showed that measurements of the direct and mirror-image reflections were equivalent to having a dual-Doppler configuration with the second virtual radar located at the mirror-image position of the aircraft below the ocean surface. This then permits the determination of both the cross-track component of the wind and the vertical motion of the precipitation. Moreover, the use of the mean fall speed of the rain and its reflectivity could provide two variables for improved estimates of rainfall. This work was later extended in a study to demonstrate the possibility of simultaneous measurement of ocean radar cross section and rainfall rate by a nadir-looking radar in space (113). Such mirror images have now been seen regularly by the TRMM radar. Once more this reemphasizes the serendipity lesson: Be prepared for the unexpected, for therein lies the source of a new discovery.

During the latter study, I wondered why we had not seen such mirror images in prior airborne radar observations. I recalled the observations made by Lou Battan with a vertically scanning radar in studies of precipitation over the Caribbean Sea in the early 1950s. So, I went back to Lou's first book on radar meteorology and there, facing the title page, was a picture of stratiform rain echoes with mirror images extending below the sea surface (114). For three decades, none of us had thought that these echoes were more than artifacts. Lou was appropriately surprised and pleased. Of course, the key reason that we were able to detect the mirror images so easily on the ELDORA radar was that the display was in color and represented the well-understood motions of the precipitation. It is also interesting to note how older scientists can retrieve long-forgotten work from their memory banks.

During the review process of the first mirror-image paper, the editor suggested deleting the story of how we first detected this phenomenon. I believe that such anecdotes are important elements of scientific reports and should be published, both for instructional and historic purposes.

Finally, in the interim between my resignation as laboratory director and retirement from NASA, I was a member of the System Z team that undertook a far-ranging study of the needs for, and means of conducting, a long-term comprehensive measurement program of the climate system. Our reports were ultimately revised and whittled

down to form the basis for the Earth Observing System. This was a
most enjoyable activity because it involved interactions with virtually
all those who had working or proposed methods for remote sensing of
parameters of the oceans, land, and atmosphere. It was another extraordinary learning experience to end my formal NASA tenure and
start a third scientific career.

My retirement party took place on Saturday, 17 November 1984 at
the University of Maryland Center for Adult Education. It was a much
grander affair than I expected with family, friends, and colleagues,
some who had come from considerable distances, and dancing to a
ten-piece band. To my embarrassment, the organizers had also solic-
ited contributions for an expensive gift from friends all over the
country and abroad. I was greatly moved by the tributes, which were
more like eulogies marked by exaggerated flattery. But I joined in the
hearty laughter at the roasts; the speakers would not let me forget my
failings and sometimes overzealous missteps.

Surely, the most lasting words were those that came in the form of
a memory book of letters from friends everywhere. As I read those
letters today, some 17 years later, I choke up with emotion at the
anecdotes that piece together much of my professional life and the
friendships I've made along the way. These memorabilia are also prone
to excessive flattery, but I am inclined to believe that there are grains
of truth sprinkled throughout. Of course, there were no letters from
those of my associates whom I may have hurt by my impatience and
occasional impromptu criticism. I prize all these letters deeply. None
have meant more to me than those from Joanne Simpson and Bob
Serafin. I have chosen not to include them here because they are
excessive in their praise, but they reflect our shared excitement in
discovery and research philosophy, our mutual admiration, and our
warm personal relationships. I can keep them to myself and read them
from time to time as therapy when I get low.

A couple of paragraphs from my farewell response at the party
follow.

> Each step [in my career] was orgasmic and euphoric, feelings to which
> one becomes addicted. Science then becomes an obsession that drives
> you on. The joys of success are exquisite; the pains of failure and
> frustration, excruciating. I do not know a scientist worth his salt who
> is bland and blasé about his work. I admit that my enthusiasm is

sometimes excessive (as is my impatience with bureaucratic obstacles) but that is the source of both my weakness and my strength.

When in either professorial or managerial roles, the nature of one's pleasures takes on other tones. Your joys come increasingly from the discoveries of others, the pride in seeing your students, protégés, and staff mature and take their place at the forefront, and in the impact which your laboratory or institution is having on the discipline. I have been lucky to have had an abundance of this kind of satisfaction in every one of my positions.

One of those who have given me cause for immeasurable pride is Bob Serafin, whom I advised during his doctoral work at the Illinois Institute of Technology (IIT) and brought to NCAR as head of the Field Observing Facility. As noted earlier he subsequently became director of the Atmospheric Technology Division and NCAR director. He resigned in the spring of 2000 after serving for more than a decade, the longest tenure of any of the directors. And in January of 2001, he took office as president of the American Meteorological Society. His position as NCAR director and now as AMS president is a singular recognition of his wisdom and judgment and the high regard with which he is held in the community. Indeed, he is the only engineer to have occupied both these positions. We have enjoyed a remarkable professional relationship and an enduring friendship, shared with our wives, for more than three decades.

CHAPTER 9

The Wanderer, 1984–2000

My official retirement from NASA was dated 3 December 1984. On that day, I started employment as a distinguished visiting scientist at the Jet Propulsion Laboratory, California Institute of Technology. That was the beginning of visits to Pasadena for one to two months each year. During the period from April 1985 to July 1987 I was also a senior research associate at the Department of Meteorology, University of Maryland, where I had grants from both the NSF and NASA. Lucille and I started to spend three to four months in southern Florida, starting in 1989, when I also worked part-time with Frank Marks at the Hurricane Research Division of the NOAA Atlantic Oceanographic and Meteorological Laboratory (AOML) and the Cooperative Institute for Marine and Atmospheric Sciences (CIMAS) at the University of Miami. The period 1986–1990 was also devoted in large part to the planning and organization of the Battan Memorial and 40th Anniversary Radar Meteorology Conference, and the editing of the resulting book *Radar in Meteorology*. During the entire period, I retained an office at Goddard Space Flight Center.

Section 9.1. The Jet Propulsion Laboratory

My initial activities at the Jet Propulsion Laboratory (JPL) were with Eastwood Im, Fuk Li, Bill Wilson, and Steve Durden in the Radar Division where we worked on the design of the airborne rain mapping radar (ARMAR) and a conceptual design of a spaceborne cloud radar. Im and I also did some work on rain retrieval from a spaceborne radar while Meneghini and I continued our studies of rain retrieval algorithms at Goddard.

However, the most exciting results of my visits to JPL concerned footprints of storms on the ocean as seen by synthetic aperture radar (SAR). I spent a great deal of time in discussions with Ben Holt, who had done studies of ocean backscatter as seen by the SAR on the SEASAT satellite. Fu and Holt had produced an impressive report describing the views of the oceans and sea ice from the synthetic aperture radar on board SEASAT (116). This comprised a collection of SAR images along with initial interpretations of the origin of the patterns. One of the images in that collection resembled storm echoes seen by conventional radar, and the caption attributed them to such storms. The question was, how did such echoes actually arise? This intriguing picture troubled me for several years. The 23 cm radar was not sufficiently sensitive to detect the rain directly. Rather, the SAR echoes were due to Bragg scatter from 30-cm wavelength capillary-gravity waves on the sea, with total echo areas of 20 to 30 km. Within the core of the echo area was an echo-free hole a few kilometers in size. The investigation turned out to be a fascinating detective story that took me far afield from meteorology. Ultimately I hypothesized that the echo-free hole was due to the damping of the capillary waves by the heavy rain within the storm core—as had been observed by Osborne Reynolds in 1875 and 1900, and demonstrated by a variety of laboratory experiments in modern times. While the heavy rain in the storm core damped the waves there, the associated downdrafts diverged in all directions upon reaching the surface. The diverging winds then produced the capillary waves that were detected by the SAR. In short, the SAR images were consistent with the structure of a downburst and outflow region produced by a convective storm (116).

A more dramatic SAR image from the ERS-1 satellite was provided to me by colleagues at the European Space Agency and Johns Hopkins Applied Physics Laboratory. The latter SAR operated at a wavelength of 5.6 cm so that the Bragg wavelength at the 23-degree nadir angle is only 10 cm. In this case, the storm occurred off the Atlantic Coast within range of the weather radar at Cape Hatteras, thus providing greater confidence in the mechanism described above. In addition, the storm in question did not move. Thus one could detect the wind-generated capillary waves and sea-surface echoes extending out in the form of a plume in a direction corresponding to that of the winds carried down from aloft in the evaporatively cooled downburst (117). I did several other papers on this subject with Peter Black of the Hurricane Research Division (NOAA AOML) and with Toshio Oguchi of the Japanese Communications Research Laboratory. The ocean-surface SAR imagery is particularly tantalizing because it provides insights on atmospheric processes that would otherwise have gone undetected.

A sidebar to the latter work is my own experience when we

encountered a moderate rain shower while sailing on Buzzards Bay at Cape Cod. Prior to the rain, the waves were about 1 foot in amplitude and one could readily see the capillary waves on the surface. Within some 10 minutes the capillary waves disappeared, the waves died down, and all that remained were the raindrop stalks generated by the bouncing raindrops and the surrounding ring waves spreading from the points of impact.

Section 9.2. Tropical Rainfall Measuring Mission

Because I retained an office at Goddard Space Flight Center and did not have significant interactions with any of the staff at the University of Maryland, I gave up my association with the university at the end of my NSF grant in 1987. However, rather than affiliate with one of the existing contractors working with Goddard, I decided to go it alone. For bureaucratic reasons, I needed a company name. I chose "Atlas Concepts" based upon a cartoon I had seen. Had not the name "Eureka," signifying "I have found it," been preempted by a vacuum cleaner company, I would have chosen that, since it characterized my excitement every time I gained a new insight. I then spent my time at Goddard where I had daily contact with my cohorts working on TRMM. In 1988, I was appointed Distinguished Visiting Scientist at Goddard so that I could also enjoy the perquisites of a NASA association.

The basic goal of TRMM was the measurement of rainfall over the Tropics of the globe from space and, for purposes of validation, from the ground. I thus returned to my first love—radar rainfall measurements. In the early 1980s Doneaud and colleagues (118) had found that one could estimate the average rainfall from a convective storm by the average lifetime area (area time integral or ATI) of its radar echo within a specified threshold rain rate or corresponding radar reflectivity. Their results were surprisingly good, but I remained skeptical, because there was no physical explanation of the method. Chiu (119) had also found that one could estimate the instantaneous area-wide rainfall from a number of storm cells by the fractional area covered by radar echoes exceeding a preset threshold rain rate. (Both approaches were later named the *threshold method.*)

These surprising results intrigued us so that we went in search of the origin of such behavior. This seemed to be related to the work of

Calheiros and Zawadzki (120) on the *probability matching method*
(PMM), in which one could determine the relationship between radar
reflectivity and rainfall rate by matching the values at which their
cumulative probabilities were equal. This suggested that the ATI
approach must be based upon the probability distribution of the rain
rate. Indeed, this was also signaled by the earlier remarks of Geof
Austin and Sean Lovejoy at McGill University when they noted that
one could estimate the storm rainfall rate simply by measuring the
area of a storm and using the known climatological rain rate for such
storms. It was then a matter of developing the simple theory and
crunching a large set of numbers for each of a set of geographical areas
to demonstrate the basis for the ATI or threshold method (121, 122).

Although the basic concept is indeed simple in retrospect, Danny
Rosenfeld (a postdoc from Israel), Dave Short, and I took special
pleasure in this particular revelation, probably because it was staring
us in the face for so many years. In Part II (122) of this work, Danny
added the measurement of storm height, which could be readily measured by radar, as an additional parameter—i.e., the taller the storm,
the greater the mean rain rate. Thus there emerged the acronym
HART for the height–area rainfall threshold method. Of course, there
are many subtleties involved in these approaches, and so there have
been literally dozens of papers by Kedem, Short, and others to improve
and elaborate the threshold method.

One of the problems with the threshold method was the need to
ascertain the actual rainfall rate R to which the radar reflectivity Z
threshold corresponded. Thus, we once again had to determine the
appropriate Z–R relationship. This then led to a series of papers on the
probability matching method (PMM) (123) by which one could obtain
the radar–rain relationship and its systematic variation with the type
of precipitation and radar range. Incidentally, we also showed (124)
that the GOES precipitation index (GPI) of Richards and Arkin (125),
a measurement of the area encompassed within a specified IR cloud-
top temperature that is related to average rainfall rate at the surface,
is also an ATI method. And we used the PMM along with airborne
radar and rainfall measurements to determine the radar–rain relationship in hurricanes (126). I refer you to the review paper in which
many of these developments are discussed (127).

An illuminating incident connected with TRMM occurred when
NASA headquarters suddenly cut $15 million from the mission budget

at a critical time. Joanne Simpson was frantic and wanted to contact someone in Congress. However, I was the only non–civil-service person who had some influence and could lobby directly. So, I called the office of Senator Barbara Mikulski and asked for her science specialist. To my pleasant surprise, it was Tom Spence, a former student from the University of Chicago who was on assignment to the senator from NSF. I explained the probable impact of such a cut at the time. At his suggestion, I prepared a one-page impact statement. Within a few days, the funds were restored. In the case of big space programs, politics is as important as science.

Danny Rosenfeld, a postdoc from Israel, warrants special mention. He is a bundle of energy and a boundless source of good ideas. We teamed up in a series of papers on the threshold and PMM methods, as noted above. His energy was infectious and his ideas stimulating, so that he carried me along by his momentum. He quickly made a major impact upon the methods of ground validation of rainfall for TRMM. He participated in a number of workshops on baseline ground truth methods; although his approach was not adopted, he exerted a strong influence on the scheme that was chosen. Upon the death of his professor and my friend, Abraham Gagin, Danny returned to Israel to head the Cloud and Precipitation Physics Laboratory at Hebrew University. There he returned to his first love of weather modification, initiating collaborative experiments in Thailand and Texas. When TRMM was launched, he combined the observations from TRMM and other satellites to demonstrate that smoke and polluted effluents from major industrial plants were inhibiting the occurrence of rainfall over vast areas. Also, he has devised methods to identify those clouds that are likely to respond positively to cloud seeding. His work on the inadvertent decrease of precipitation by man's activities has taken on an importance that appears comparable to that of the effect of greenhouse gases on global warming and climate change.

The 1996–2000 period was dominated by studies of the nature of the drop-size distributions (DSD) in tropical rains and the relationships between radar reflectivity and rain rate, undertaken with my long-time cohort, Carl Ulbrich, and Frank Marks of the Hurricane Research Division of NOAA/AOML. The latter studies are particularly relevant to TRMM because of the need to distinguish stratiform from convective rains and to develop algorithms to measure rain from both space and the ground. We used surface data taken with a Joss-

Waldvogel disdrometer on Kapingamarangi Atoll in the equatorial
Pacific and from an airborne drop size probe on the NOAA P-3 aircraft
during the Tropical Ocean and Global Atmosphere Coupled Ocean-
Atmosphere Response Experiment (TOGA COARE) (128). This work
yielded a set of four interesting papers, of which I will discuss only two.

The first was the finding that the characteristics of the tropical
rainfall vary systematically with the rain type and its evolution in
space and time (129). For example, the first drops from a convective
shower tend to be large, uniform in size, and steady for periods of
20–30 minutes, while the rain rates vary sharply. This is what we all
observe in the first rains of summer showers. The nearly constant large
size during this period means that the radar reflectivity Z is directly
proportional to the rain rate R, contrary to conventional wisdom. The
initial convective rain is followed by a transition period during which
both the drop size and rain rate decrease sharply, and then by a period
of more or less constant stratiform rain at rates less than 10 mm h^{-1}
for an hour or more. Prior findings erroneously assigned the radar–
rain relation of the transition period to the convective rain because the
two types were not separated from one another. Our interpretation
was aided by the use of a vertically pointing radar profiler and the
collaboration of Christopher Williams of the NOAA Aeronomy Laboratory in Boulder, Colorado.

The final work on tropical rain proposed a physical mechanism
responsible for the behavior of the DSDs measured aloft by the aircraft
and those observed at the surface during convective rains (130). We
found that the DSDs observed aloft during rains greater than about 20
mm h^{-1} resembled so-called "equilibrium" size distributions, i.e., those
that obtain a steady-state form resulting from the combination of drop
growth and breakup by collisions. To our pleasant surprise, we also
found that the drop size at which the rate of collisions is optimum has
a fall speed that matches the updraft speed measured by the aircraft.
Thus, these drops are nearly balanced, while much larger ones fall out
below the updraft core and much smaller ones rise and fall out
elsewhere if they survive. The updraft thus performs two functions: (1)
it supports the middle range of drop sizes for an extended period of
time to allow the collision process to produce an equilibrium DSD, and
(2) it separates the smaller drops from the large ones so that only the
latter reach the surface. This explains our prior findings of nearly

uniform large drop sizes during the initial convective rain in the previous paper.

I'd like to pause here to say a few words about Prof. Carlton Ulbrich, who has been my collaborator and friend for 28 years. As noted earlier, we first joined forces when he came to the Advanced Study Program at NCAR in 1973, again when he worked with us on the Hail program there in 1974–75, and even more frequently since I moved to NASA in 1977. Our research dealt with almost every aspect of rain (and hail): its measurement by in situ and remote sensors, its effect upon microwave communications, and recently the physical processes controlling the nature of the drop-size spectra. He has also studied lightning using the giant radar at Arecibo in Puerto Rico. Carl is a top-notch physicist and mathematician. And he can make a computer sing. He is a delight to work with; nothing seems beyond his grasp. A question posed in the morning will be answered in the afternoon by e-mail, often with intricate but readily understood illustrations. He is usually patient with me, but is not beyond showing annoyance at either my excessive demands or my naiveté. He is one of the dozen colleagues with whom I have enjoyed that beautiful symbiotic and explosively productive relationship. I cherish him as a friend and colleague.

While I am at it, I must say something about Frank Marks at HRD in Miami. Frank is in charge of the aircraft hurricane research activities. He knows about as much about tropical storms as anyone in the world and is equally expert on the use of Doppler radar and other remote and in situ sensors. Moreover, he is as cooperative as they come. He is a one-man national resource who has an international network of friends and collaborators. He too is a pleasure to work with, and I am indebted to him for countless favors.

CHAPTER 10

The Battan Memorial and 40th Anniversary Conference on Radar Meteorology and the Conference Book *Radar in Meteorology*

The first conference on radar meteorology was a small informal meeting held at MIT on 14 March 1947. We did not know at that time that there would be others, so it was not known as the "first." By 1986, we had held 23 conferences spaced about 21 months apart on average. Rogers and Smith (2) provide an interesting commentary on the important role of these conferences in stimulating progress in the field and setting the style for AMS meetings in all other specialties.

As one of the senior radar meteorologists and one of only three other survivors who attended the first conference, I thought it would be appropriate to celebrate the 40th anniversary of that meeting with the goal of taking stock and setting an agenda for the future. Rit Carbone, then chairman of the AMS Committee on Radar Meteorology, agreed, but only if I would assume the role of chairman and editor of the proceedings. Ralph Donaldson, my friend and colleague from AFCRL days, agreed to cochair.

Prof. Louis (Lou) Battan, one of the pioneers in the field and a prolific scientist and author, and my WWII roommate and cherished friend, died on October 29, 1986, shortly after the planning had begun. The conference and the resulting book *Radar in Meteorology* (3), based upon the conference, were therefore dedicated to Lou's life and memory. My tribute to Lou appears in the book. That tribute had previously appeared in *EOS*, the weekly newspaper of the American Geophysical Union. When Jeannette Battan received her copy (in August 1987), she responded with a touching letter, which I treasure.

Radar in Meteorology is a unique volume. It combines (1) a set of histories of the early evolution of the field in various institutions and nations; (2) 10 essays on the state of the art of the technology and the scientific topics to which radar has contributed, and supplementary commentaries; and (3) three reviews of the operational applications of radar to the measurement of precipitation, severe storm detection, and aviation. Most of the essays were prepared subsequent to workshops on the subject held in advance of the actual conference; thus they reflect a general consensus of the community.

The editing of this massive volume was a tour de force and labor of love. I devoted the better part of two years to the project. In spite of the exhausting effort, I enjoyed every moment. In 1989, I spent a week in the hospital with an attack of phlebitis. Bored by the long days in bed, I spent much of the time editing, focusing upon Chapter 27a—"Radar Research on the Atmospheric Boundary Layer"—by Earl Gossard. This 50-page chapter is worthy of a book in its own right. It exhilarated me so that I had to call Earl from my hospital bed to express my excitement and compliments. Some chapters, like Earl's, were so well written that they required little editing. Others presented controversial remarks and weak hypotheses that required intricate negotiation with the authors and revisions. At the end of the day, the book exceeded my fondest dreams. I shall be forever grateful to the many people who collaborated in this endeavor.

CHAPTER 11

Radar Detection of Low-Level Wind Shear

Another chapter of my life after NASA concerns the invention and patent on a method of using the conventional airport surveillance radar (ASR-9) to detect hazardous low-level wind shear (131). Prior to mid-1988, wind shear had been responsible for twelve major aircraft accidents, seven of which resulted in the loss of 575 lives. Thus, wind shear detection and warning was high on the FAA's priority list of critical problems. Westinghouse Electric, the company responsible for the development of the ASR-9, called me in to discuss the possibility of upgrading the radar to include a wind shear detection capability. I was impressed with the list of distinguished company radar experts at our first meeting on September 19, 1985. But it soon became clear that they had not the slightest idea of how one might approach the problem. Neither did I at that point. The essential obstacle was the fact that the antenna radiated two wide vertical fan beams, one tilted slightly above the other. These beams would therefore detect all the precipitation over a broad altitude range, whereas it was necessary to measure only the Doppler velocities and wind shear at the lower levels where the aircraft were flying during takeoff and landing. After several hours of discussion, none of us had any concept of how to proceed. Nevertheless, I agreed to think about the problem and get back to them later to sign a consulting agreement.

On the drive home from Baltimore, the idea struck. Eureka! I thought of a method by which we could get the low-level winds. The essence was to use the difference in the Doppler spectra (i.e., the echo power returned at each Doppler velocity) between the two beams so that the velocities at one bound of the difference spectrum corresponded to the near-surface winds. I immediately dropped everything else and started to write at a phenomenal pace. As usual with such epiphanies, I was so excited that I lay awake much of the night working out the details. Meanwhile, I called Westinghouse to tell them that I could not consult for them, and retained a patent attorney to prepare a patent application. I wrote nonstop, completed an invention disclosure on September 27, and had it witnessed by my colleagues by October 3. I worked intensively with my attorney on the detailed description, and most importantly, on the 47 claims. It was a remarkable achievement to file the patent on November 8, 1985. (The official filing date was later changed to December 23, 1985.) The patent was issued on March 10, 1987. I also filed an application for reissue on December 5, 1988 to broaden the claims; Reissue Patent 33,152 was granted on Jan 23, 1990.

Meanwhile, I arranged a presentation to Neil Blake, the director of engineering for FAA, on November 15, 1985. In addition to a couple of FAA staff members, Blake had called in two weather radar experts, the leaders of the MIT Lincoln Laboratory Weather Sensing Group concerned with aviation weather problems. Both were long-time colleagues. It was the most frustrating presentation in my career because Expert #1 kept shaking his head and interrupting me to say that the method would not work for one reason or another. I had encountered criticism frequently over the years, but never before had I been so enraged. Nevertheless, I restrained myself. At the end of my presentation, Neil Blake tossed my unopened disclosure back to me saying that he concurred with his "experts" and abruptly ended the meeting.

The plot thickened when I received a 1987 Lincoln Lab report by Weber and Moser (132) in which they proposed a dual-beam method for wind shear detection that was virtually identical to my system. This was most suspicious because, as was later documented, although they were working on aviation safety problems by radar, they had no idea of using a dual-beam Doppler difference scheme prior to my disclosure (John Anderson, personal communication, Nov. 28, 1988). Moreover, the Weber-Moser report simply acknowledged my patent in a legalistic footnote to the effect that "David Atlas had developed a similar method" rather than in the proper scientific manner. Dr. Walter Morrow, then director of Lincoln Lab, subsequently confirmed my priority on the method. In conversations with Mark Weber, he apologized for improper acknowledgement of my work and acknowledged that the footnote was the work of their attorneys. We have maintained a friendly relationship over the 15 years of this sordid affair as Lincoln Lab proceeded to develop and successfully field-test a dual-beam prototype system (133). Later still, Weber developed and patented an autocovariance method for processing and deriving the wind shear signals (134). It is the latter approach that is the basis of the FAA claim that their system does not infringe my patent.

This is the most egregious case of plagiarism that I have encountered in the 56 years of my career. Expert #1, the leader of the group responsible for developing hazard detection methods for the FAA, made the unbelievable claim that he never communicated my method to his team. Rather it was a simple case of nearly simultaneous independent discovery. Whether or not he divulged my scheme explicitly, ideas such as these pass from one to another rapidly during supposedly innocent conversations. This is one of the ways in which science advances. But ethical behavior requires appropriate acknowledgement of prior art. Perhaps this incident hurt so much because it was the theft of one's brainchild, which scientists hold so dear.

After years of foot-dragging by the FAA, Northrop-Grumman (who took over Westinghouse Electric) developed and deployed five prototype ASR-9 radars with wind shear processors in 1999–2000. An additional 32 systems are being installed during 2001. At this writing, an infringement suit is in abeyance. In spite of the obvious friction that has accompanied this episode, I am happy to note that Mark Weber and I remain friends. He is a delightful person, a clever scientist/ engineer, and a perfect gentleman.

CHAPTER 12

Some Thoughts on Science Management

Here I shall indulge myself with a bit of scientific philosophy. I do not take credit for these ideas, for they come from those who have devoted far more thought to the subject than I. However, I have assembled a few of the more profound concepts in a paper that I presented as part of the first remote sensing lecture of the AMS in Paris in 1991 (135). Perhaps the entire article will convey the intensity of my feelings for the subject, but the following selections stand on their own.

Thomas (105) has described the research process eloquently:

In basic research, everything is just the opposite [to that in applied]. What you need at the outset is a high degree of uncertainty; otherwise it isn't likely to be an important problem. You start with an incomplete roster of facts, characterized by their ambiguity; often the problem consists of discovering the connections between unrelated pieces of information. You must plan experiments on the basis of probability, even bare possibility, rather than certainty. If an experiment turns out precisely as predicted, this can be very nice, but it is only a great event if at the same time it is a surprise. You can measure the quality of the work by the intensity of astonishment. The surprise can be because it did turn out as predicted, or it can be confoundment because the prediction was wrong and something totally unexpected turned up, changing the look of the problem and requiring a new kind of protocol. Either way, you win.

What a magnificent description of the research process!

As Gilman (136) points out, the very words "research" and "development" are like oil and water; they don't mix. To be sure, the ecology and management of applied science and development need to be different from those of basic research. Unfortunately, the two are often commingled so that they go on side by side. Managers, whether in industry or government, are also inclined toward mission orientation because, according to Homer Newell (104), " . . . to invest funds in support of research that holds greatest promise of a specific desired

117

application is the most easily justified and patently wise course of action."

And since management inclines toward clearly defined procedures, with well-defined targets, approaches, plans, and schedules, they tend to go by the book. But the rulebook is inimical to the basic scientist; it constrains him along predefined tracks, and precludes him from going in hot pursuit of the unexpected, the surprise that requires that he deviate from the project plan.

There is no cookbook for creativity; the targets are elusive, the search is broad, risky, and fraught with traps; and serendipity plays a key role in discovery. Indeed, if the lucky break did not mark creativity, science would be a boring routine.

Dalrymple (137), in his plea for a balance between big and small science, emphasizes that the major portion of fundamental discoveries has come from small and undirected research. The operative word here is "undirected." Moreover, he notes that the unique feature of undirected science that makes it essential to overall progress in science and technology is diversity—diversity because neither scientists nor administrators are able to predict the variety of useful products that may blossom from basic research. This kind of diversity is intrinsic to a kind of Darwinian evolution, first, because we know not from which mutant seed the surviving concept will spring, and second, because we are unable to predict the utility of the surviving variants.

I suggest that the above mentioned papers be required reading for every science program manager. I hope that some of you will read my original missive because it also speaks to the role of the "stars" in the creative firmament in passing on their intellectual genes to those with whom they interact, and the need to infuse new blood into the institution when the stars pass on.

CHAPTER 13

Retrospective

When the experiences of 56 years are compressed into some 150 pages, one sees his life in accelerated time lapse. I find it difficult to conceive of all that has been packed into those years—the various positions I've held, the wide range of research that we have undertaken, the thrills of discovery, the pure wonder of how it all comes together, the excitement of telling the world about it, the heartwarming recognition by one's peers, the sweetness of working with inspiring colleagues, the pleasures of seeing your protégés make their own mark in the world of science, and making friends around the globe. What exquisite joy! What a magical adventure! The frustrations of fighting the bureaucracy and having one's proposals and manuscripts rejected from time to time pale by comparison.

The seemingly random activities of the last half century have been integrated into a sensible picture, largely because we build upon each other's advances, and partly because we don't let each other proceed along fruitless lines for very long. With all its problems, the peer review process works, either formally through the funding agencies and journals, or informally, through interactions with one's colleagues locally and at conferences. Peer review does not work well when funds are scarce or when two groups are racing for priority in discovery. Then the competing investigators have obvious conflicts of interest and should not be reviewers. At times like this, one needs to have more objective reviewers who may not be as expert in the subject matter.

It is ironic to think that the death and devastation of WWII yielded so much good. Wars have stimulated scientific and technological progress throughout history, but one can hardly balance the benefits of WWII against the unprecedented tragic loss of life in that war. Radar is a two-edged sword. On the one hand, it continues to be a phenomenal weapon; on the other, it has yielded innumerable scientific advances and societal benefits. And those of us who have worked in meteorology, oceanography, and aviation have helped bring them to fruition.

I have been particularly well blessed with good fortune as a researcher. First, it was our wartime training in radar that provided us with the combination of a telescope and the X-ray to view the atmosphere in entirely novel ways. This in itself was an accident. Weather was noise to the detection and warning of enemy activity, but it was the signal to the meteorologist. One cannot imagine the excitement and wonder of seeing the first images of storms, precipitation trails, the bright band, or hurricanes and tornadoes. Nor could we believe for many years that we could see magnificent structures in the clear air. We were there at the very dawn of the age of remote sensing. Since then, I have urged students to make themselves unique by combining two disciplines, for major advances are often found at the interfaces of the two.

But there was much more. It was also partly fortuitous to find myself at the right place at the right time with the right management. When things went wrong, as they did in the hail program at NCAR, I moved on. It was the luck of the draw that I was assigned to the All Weather Flying Division after radar school in 1945. Also, what could have been better than to become associated with the Thunderstorm Project through my boss, Major Joe Fletcher? Who would have guessed that a hurricane and I would meet in Orlando where I had access to the most advanced radars in September 1945? All I had to do was to get hold of a camera, film, and food enough to keep me going for 36 hours.

My colleagues and I have been able to exploit the combination of flexibility, serendipity, and opportunity that one finds in a hospitable environment. With rare exceptions, our management was supportive, both morally and financially. We had the freedom to go in hot pursuit of the unexpected, which implies that we had some form of hypothesis to start with and were alert to surprises—the essence of serendipity. And we made opportunities by forging associations with colleagues wherever the conditions and instruments were better adapted to the target weather phenomena. Our takeover of the soon-to-be-closed major radar facilities at Wallops Island was just one of many such examples. The story I've told is replete with many more.

Conferences and exchange visits with other institutions were also vital ingredients. At AFCRL and the University of Chicago, we would close up shop and take most of the staff, including technicians, to the radar meteorology conferences. They then learned about what was going on elsewhere and saw how our research meshed or clashed with that of other groups. They also gained a sense of pride in our own group. Our continuing relationship with the McGill Stormy Weather

Group was intrinsic to our progress and to theirs. We have recounted the impact of the long series of radar meteorology conferences on the evolution of the field, and of the series of meetings of the Inter-Union Committee on Radio Meteorology on the discoveries of the previously inconceivable fine-scale structure of the clear-air atmosphere. And we planted seeds around the world through the series of visitors who joined us at AFCRL and the University of Chicago. I doubt that Doppler radar would have advanced nearly as rapidly had we not brought Roger Lhermitte to this country. But there were many other visitors who nourished us and the discipline. Among these, I must mention Milton Kerker, who began his book on particulate scattering while with us. Others who must be acknowledged are Jiro Aoyagi and Keikichi Naito of the Japanese Meteorological Research Institute, Ryozo Tatehira of the Japan Meteorological Agency, Jurg Joss of the Swiss Institute of Meteorology, and Hans Ottersten of the Swedish Defense Research Establishment. Nor can I ignore the vast influence of Dag Gjessing (formerly director of the Norwegian Remote Sensing Technology Program), whose imagination leaped far beyond extant knowledge.

One cannot measure the impact that my sabbatical year in England had on me, or upon the subsequent evolution of radar meteorology there. The story of the Wokingham storm has been told several times. I returned to the States with a new philosophy inspired by Frank Ludlam and Keith Browning. But subsequent events are even more important. Keith Browning revolutionized our understanding of severe storms during his four-year stay in my laboratory at AFCRL, and went back to the United Kingdom to reinvigorate the virtually dormant field there. He remains a one-man powerhouse. I was happy to join him, John Goddard, and Anthony Illingworth in supporting the continued operation of the world's largest multiparameter weather radar at Chilbolton (operated by the Rutherford Appelton Laboratories) whenever it came under budgetary attack. Keith subsequently went on to establish the Joint Center for Mesoscale Meteorology (JCMM) at the University of Reading and the Universities Weather Research Network (UWERN), a consortium of U.K. universities. Recently he led a UWERN bid that has been successful in winning support for the procurement of major observing facilities to be based at six universities, but operated collectively as the (UK) Facilities for Atmospheric Monitoring (UFAM). Nor must we forget his profound influence upon international programs such as the World Climate Research Program (WCRP) and the Global Energy and Water Experiment (GEWEX).

Much of what I have said about Keith Browning also applies to all of the remarkable scientists whom I have mentioned in this essay—Stewart Marshall, Walter Hitschfeld, Raymond Wexler, Milton Kerker, Frank Ludlam, Roger Lhermitte, Ramesh Srivastava, Carl Ulbrich, Bob Meneghini, Joanne Simpson, and Danny Rosenfeld, among others. They have all worked at the cutting edge of the discipline and have set the stage for successive generations. Who knows where it really started and where it will end? To recoin a phrase, "a teacher's influence never ends." I hope that I shall be numbered among those teachers.

Epilogue

There is so much more that I am bursting to tell, but we must bring this story to a close. I continue to delight in the host of advances by young scientists around the world and feel sure that the years ahead will be as exciting as those that have passed. As I scan the program of the 30th Radar Meteorology Conference held in Munich in July 2001, the first in 54 years that I did not attend, I feel a sense of loss. I regret not having been there to listen, to learn, and to kibitz.

However, as I write this farewell message in the spring of 2001, it appears that I shall not have to endure the shock of a sudden departure. Rather, I find myself happily entrained once more in research. The collection of radar, aircraft, and rainfall observations from a field campaign in Brazil again promise to reveal the inner workings of tropical convective storms. Once I saw the awesome observations of their structure as seen by the Christopher Williams' (NOAA) radar profiler I could not resist the temptation to join the search. Nor could I resist the invitation to join the NAS Committee on Weather Radar beyond NEXRAD as one more opportunity to influence the directions of radar meteorology in the twenty-first century.

Barring a few ailments, it is my good fortune to remain in reasonable health. Good fortune has carried me from the dawn of radar in mid-twentieth century to a rebirth of the field at the start of the new millennium. The Tropical Rainfall Measuring Mission (TRMM) has been a crowning achievement, with advances and applications that we never imagined when it was first proposed. I came away from the TRMM science meeting at the end of October 2000 with a feeling that our vision is myopic: We can only see a short distance into the future. But TRMM, its successors, and its sister satellites are bound to have as great an impact upon the earth sciences and applications as did Galileo's use of the telescope on our understanding of the cosmos. This is not the end of the story; it is just the beginning.

Acknowledgments

I hope that I have expressed adequate appreciation to the many colleagues who have made my career in meteorology such a joy, and that I will be forgiven for having missed a few. Here I wish to recognize those who were pivotal in motivating me to write this story and who have provided significant commentary and editing. They include Joanne Simpson, Ralph Donaldson, Rod Rogers, Carl Ulbrich, Bob Fleagle, Linda and Rit Carbone, Bob Serafin, Bill Bandeen, and my wife Lucille. Akiva Yaglom and Valerian Tatarski also cleared my fuzzy recall of our visits to the Soviet Union in 1965 and 1971. My memory of events has also been jogged by many others to whom I refer in the text. Special appreciation goes to John Perry, the AMS editor for historical monographs, who combines a great sense of meteorological history with his writing skills, and to Ken Heideman, AMS Director of Publications for enthusiastic encouragement and guidance. I add my special appreciation to Debbi McLean and James O'Leary of the Technical Information Services Branch at Goddard Space Flight Center for their imaginative work on the dust jacket. My heartfelt thanks go out to all of you.

Appendix A: Acronyms

AFCRL   Air Force Cambridge Research Laboratories
AOIPS   Atmospheric and Oceanic Information Processing System
AOML   Atlantic Oceanographic and Meteorological Laboratory, NOAA
ARMAR   airborne rain mapping radar
ASR   Airport Surveillance Radar
ATD   Atmospheric Technology Division
ATI   area-time integral
AWFD   All Weather Flying Division
CAT   clear-air turbulence
CIMAS   Cooperative Institute for Marine and Atmospheric Sciences
COLA   Center for Oceans, Land, and Atmosphere
COSS   climate observing system study
CRL   Communications Research Laboratory
DOGS   Department of Geophysical Sciences
DSD   drop-size distribution
ELDORA   Electra Doppler Radar
ERL   Environmental Research Laboratories, NOAA
FAA   Federal Aviation Administration
FGGE   First Global GARP Experiment
FM/CW   frequency modulated/continuous wave radar
FOF   field observing facility
GARP   Global Atmospheric Research Project
GATE   GARP Atlantic Tropical Experiment
GEWEX   Global Energy and Water Experiment
GISS   Goddard Institute of Space Sciences
GLAS   Goddard Laboratory for Atmospheric Sciences
GOES   Geostationary Operational Environmental Satellite
GPI   GOES precipitation index
GRD   Geophysics Research Directorate at AFCRL
GSFC   Goddard Space Flight Center, NASA
HART   height-area rainfall threshold
HIRS   high resolution radiation sounder
HRD   Hurricane Research Division
IIT   Illinois Institute of Technology
IUCRM   Inter-Union Committee on Radio Meteorology
JCMM   Joint Center for Mesoscale Meteorology
JDOP   Joint Doppler Operational Project
JPL   Jet Propulsion Laboratory, California Institute of Technology
MASEX   mesoscale air-sea exchange

McIDAS	Man-Computer Interactive Data Analysis System
MEW	microwave early warning
MST	mesospheric-stratospheric-tropospheric
MSU	microwave sounding unit
NASA	National Aeronautics and Space Administration
NCAR	National Center for Atmospheric Research
NCEP	National Centers for Environmental Prediction
NDRB	National Defense Research Board
NESS	National Environmental Satellite Service
NEXRAD	next generation weather radar
NHRE	National Hail Research Experiment
NOAA	National Oceanic and Atmospheric Administration
NRC	National Research Council
NSF	National Science Foundation
NSSL	National Severe Storms Laboratory
PAM	portable automated mesonet
PIA	path integrated attenuation
PMM	probability matching method
PSI	plan shear indicator
RSRE	Radio and Space Research Establishment (UK)
RTOP	research and technology operating plan
SAR	synthetic aperture radar
SODAR	sound detection and ranging
SRT	surface reference technique
TOWR	Terminal Doppler Weather Radar
UAL	United Air Lines
UCAR	University Corporation for Atmospheric Research
UFAM	U.K. facilities for atmospheric monitoring
UGGI	International Union of Geodesy and Geophysics
U of C	University of Chicago
UWERN	Universities Weather Research Network
VAD	velocity-azimuth display
VAS	Visible and Infrared Spin-Scan Radiometer
WCRP	World Climate Research Program

Appendix B: Biographical Sketch

David Atlas studied meteorology at New York University (B.Sc., 1946) and radar at the Harvard-MIT Radar School during World War II. He received his M.Sc. (1951) and D.Sc. (1955) from MIT while working at the Air Force Cambridge Research Laboratory (now Air Force Research Laboratory) where he led the weather radar research program for 18 years. He also taught at the University of Chicago and headed both the Atmospheric Technology Division and National Hail Research Experiment at the National Center for Atmospheric Research. He subsequently established and headed the Laboratory for Atmospheric Sciences at the NASA Goddard Space Flight Center, where he is currently a distinguished visiting scientist. He also held a similar position at the Jet Propulsion Laboratory, California Institute of Technology. He is a past president and honorary member of the American Meteorological Society. He is the recipient of a number of awards from the AMS, including its premier Carl Gustav Rossby medal, and the Symonds Memorial medal from the Royal Meteorological Society. He is a member of the National Academy of Engineering.

Appendix C: Basic Radar Meteorology

It was in the summer of 1940 in England that the first microwave (i.e., short wavelength) radar became operational. Wavelengths as short as 10 cm were made possible by the development of the magnetron. Since the backscatter from raindrops and snow smaller than the wavelength of electromagnetic radiation (so-called *Rayleigh scatter*) is inversely proportional to the fourth power of the wavelength, these short wavelengths made it possible to detect precipitation. So, it is likely that the first weather radar observations were made in England in July or August of 1940. However, the British were preoccupied with fighting the war, so relatively little effort could be devoted to the meteorological applications of radar. Thus, the early work on weather radar was done at the MIT Radiation Laboratory (Rad Lab) in the United States. It was Captain Joe Fletcher, a pilot and meteorologist working at the Rad Lab, who was among the first to appreciate the value of radar for tracking balloons in all weather and for the detection and forecasting of precipitation. He convinced the U.S. Army Air Weather Service to train 100 weather officers in radar, and thus was born "radar meteorology."

One of the most important applications of radar from the earliest days to the present is the quantitative measurement of rainfall. This is due to the fact that the radar echoes from raindrops are proportional to the sixth power of the drop diameter so that a drop of 2 mm diameter reflects 64 times more strongly than a drop of 1 mm. This is the Rayleigh scattering regime. By comparison, the rainfall rate R is approximately proportional to the fourth power of the drop diameter, so it can be shown that the radar echo power or reflectivity factor Z is roughly proportional to the second power of R.

However, the situation is even more complicated than this. Rigorously, it can be shown that Z is proportional to the total liquid water content W (which is proportional to the third power of the drop diameter) and the third power of the mean drop size by volume D_m—thus to the sixth power, as required by the Rayleigh scatter law. This means that Z depends not only upon W—which is closely related to the rain rate R—but also to D_m or the effective drop diameter. Still another complication is the dependence on the form of the drop-size distribution, but this is usually of secondary importance. In short, one can get the same rate from a large number of small drops as from a small number of large ones, but the reflectivity of the latter will be much greater. It is for this reason that modern techniques for estimating rainfall by radar now attempt to determine the effective drop size in addition to Z.

In the 1970s, it was found that raindrops tend to be approximately elliptical in shape, although the larger ones tend to have flat bottoms. Also, the ellipticity of the drops increases systematically with the drop diameter. Moreover, it is remarkable that they fall with their large axes aligned horizontally. Thus, the drops produce a larger echo to horizontally polarized radar waves than to vertically polarized waves. The ratio of the radar reflectivities at horizontal to vertical polarization is called the *differential reflectivity, Zdr*, and is a measure of the mean volume drop size D_m. And the combination of Zdr with the horizontally polarized reflectivity Z_h provides a more accurate measure of rain rate than does Z_h alone. There are a variety of methods to exploit polarized radar waves for quantitative measurements of the effective drop diameter in the radar pulse volume, i.e., the volume confined within the radar beam and the radar pulse length.

In comparison, the reflectivity of snow crystals and flakes is about one-fifth of that of a raindrop of the same mass. When snow falls through the melting level it starts to get a water coating so that its reflectivity increases fivefold. However, when it melts completely, its fall speed increases about fivefold (depending upon drop size), so that the number concentration of raindrops per unit volume decreases by a factor of five. It is only in the melting layer that the melting flakes have both the high number concentration of snow and the large echo of raindrops. Thus, we find that the wet flakes in the melting layer produce a fivefold greater reflectivity than does the snow above or the rain below. This results in the so-called *bright band* (BB) at the melting layer. The BB is characteristic of widespread stratiform precipitation in autumn through spring and on the backside of thunderstorms where the particles that have been carried up by the strong updrafts freeze and fall out as snowflakes, then melt to rain.

In thunderstorms, the strong updrafts carry the cloud droplets up rapidly. Collisions take place and the cloud drops coalesce to form raindrops; these are carried up further where they freeze to form either snow pellets (graupel) or hailstones. When either of these grows large enough to overcome the updraft strength, it falls out. Of course, the bigger ones do not melt completely; they thus reach the ground as hail. However, the frozen particles are usually covered by a coating of water even at subfreezing temperatures so they have the echo characteristics of raindrops. Moreover, since they do not change their fall speed significantly upon falling through the melting level, there is no brightband signature as in the case of stratiform precipitation. Nevertheless, the large size of hailstones causes them to scatter more strongly than drops or flakes, and so the reflectivity of thunderstorms is much stronger than that of widespread stratiform rain and snow. Of course, the stronger the updraft, the faster the rate of condensation and drop growth and the greater the rain rate that ultimately reaches the ground. In addition, thunderstorms are localized in area so that the high reflectivity is concentrated within them. They also extend to great heights, often penetrating into the stratosphere. On color-coded displays, the colors represent the intensity of the storm, as we have all seen on local or national radar weather maps. Hence, thunderstorms are easily distinguished from the widespread rain- and snowstorms that may extend over hundreds of miles or more.

Both ground-based and airborne radar have also been extremely valuable tools in mapping and tracking hurricanes and determining their intensity. The storm's eye is readily apparent, and therefore the sequence of storm positions provides a first guess as to where the storm is heading. This kind of tracking of all kinds of storms and the extrapolation of their positions for an hour or so is the basis for *nowcasting,* or very short-term forecasting.

Doppler Radar

Most of us know about Doppler radar because it is used by the police to detect speeding cars. We recognize the Doppler effect by the sound of the whistle on a moving train. When the train is approaching, the pressure peaks of the sound waves impinge upon our ears more frequently than they would if the train were standing still, so we hear a higher pitch. Conversely, we hear a lower pitch as the train recedes from us. The same is true with radar waves. The received frequency is increased (relative to that transmitted) when the targets are approaching, and vice versa. With a ground-based radar we can readily measure the speed of the precipitation particles, which move with the winds. When the radar beam is pointed either up- or downstream (with a correction for the elevation angle of the radar beam), this velocity is a good measure of the wind speed. Doppler radar is the basis for the detection and warning of tornadoes, which are identified by the presence of strong approaching and receding winds adjacent to one another in or on the fringes of an intense thunderstorm. Doppler radar has provided increased warning times of the order of 10 to 15 minutes to allow people in the paths of tornadoes to take cover. Both ground-based and airborne radars are also used very effectively for hurricane monitoring and prediction, usually in combination with satellite observations and computer models.

In the case of thunderstorms, the wind vectors usually change sharply in three dimensions. Thus, it is necessary to use dual- or tri-Doppler radar observations to map the complete wind field. A remarkable airborne radar system (ELDORA-ASTRAIA) has been developed jointly by NCAR and the French Centre de Recherches en Physique de L'Environnement Terrestre et Planetaire (CRPE), which has two back-to-back antennas rotating in a tail radome. One antenna is skewed to look forward of the normal to the path; the other looks backward. Thus, the same piece of airspace is viewed by the radar from two directions within a few minutes of flight and the wind vector can be reconstructed while the aircraft is flying in a straight line. Some of the most exciting Doppler radar measurements and images have been made with this system.

Doppler radars have also been developed to detect the wind shear and gust fronts associated with the evaporatively chilled cold-air downbursts and the outflow air from thunderstorms. These terminal Doppler weather radars (TDWR) and airport surveillance radars (ASR) provide warnings to pilots to avoid landings or takeoffs in dangerous wind shear conditions.

Wind-profiling radars use three or four beams or successive beam positions—one vertical and the others tilted slightly east or west and north or south. This combination permits the measurement of winds simultaneously at all heights at which there are detectable scatterers. The scattering elements may be either precipitation or the perturbations in refractive index in the clear air as described in the following section. The wind profilers are exceedingly valuable for providing continuous records of winds, which are used both in short-term forecasting and for the initiation of large-scale computer prediction models.

In many clear-air conditions, particularly from spring through early fall, the presence of wind-borne insects provides sufficient targets to measure winds. Indeed, such measurements are of use in a number of forecasting routines.

Clear-Air Echoes

In addition to the detection of insects, radars of 10 cm wavelength and longer are capable of detecting the small-scale fluctuations and gradients in refractive index due to temperature and humidity perturbations that occur at frontal boundaries, gust fronts, inversions, and discontinuities in wind shear. There is a very rich literature on this subject (see 28, 43). Powerful conventional pulse radars have detected the beautifully formed Kelvin-Helmholtz waves that form on inversions accompanied by wind shear. These radars have also detected the tropopause. Such waves often break to give birth to clear-air turbulence (CAT). Frequency modulated/continuous wave (FM/CW) radars provide both very high resolution and sensitivity for the detection of the fine-scale structure of the lower atmosphere. Radars and wind profilers operating at frequencies near 50, 400, and 900 MHz have been used to measure winds in the troposphere, stratosphere, and mesosphere. In this case, the scatterers responsible for the echoes are the fluctuations in refractive index at half the radar wavelength. While many other scales of turbulence are present, the radar only sums the echoes from the half-wavelength perturbations. This is called *Bragg scatter*. This kind of scatter is also responsible for the mantle echoes from the boundaries of cumulus clouds.

Synthetic Aperture Radar and Scatterometry

Synthetic aperture radar (SAR) has been used widely on board aircraft and spacecraft to map the fine-scale structure of the earth's surface over land, ice, and sea. SAR develops a large synthetic aperture by summing successive echoes from a point on the earth's surface coherently in time sequence as if the echoes from that point had been illuminated by a large antenna at one instant. The effective beam width is then as narrow as that which would be produced by a large antenna equal in width to the path length traveled by the moving platform during the integration time. Much of the earth has been mapped by a sequence of SAR systems on board a large number of spacecraft such as the U.S. SEASAT, the European ERS-1 and 2, and the Canadian RADARSAT. Similarly, both Venus and Jupiter have been mapped by orbiting SAR systems.

The author has used data from the SAR radars on board the NASA Jet Propulsion Laboratory Convair 990, SEASAT, and ERS-1. Over the sea, the echoes correspond to Bragg scatter only from the centimetric scale waves equal to half the radar wavelength divided by the sine of the nadir angle. These small capillary–gravity waves react almost immediately to the local winds. Accordingly, atmospheric processes that affect the boundary layer winds may be deduced from the structure of the wind-induced patterns as described in Chapter 9. Moreover, moderate to heavy rain damps the short waves that feed energy to the larger ones. Such damping can also be seen in SAR images of the sea below heavy rain centers. A number of authors have described a variety of SAR images over the sea that are related to local convection, to katabatic winds off the land, to storms, and to internal waves in the sea.

The relation of the amplitude of the capillary gravity waves to wind speed has also been used in radar scatterometry from space. In this case, the scatterometer takes two or more looks at the same piece of ocean from two or more directions and computes the wind vector. Recent observations with the spaceborne QuickSat scatterometer have shown closed wind circulations preceding the formation of hurricanes. The surface winds are also very important input for the prediction of large-scale weather by numerical global circulation models (GCM).

It is impossible to provide a comprehensive account of the fundamentals of radar meteorology and technology in this brief tutorial. The reader is referred to the books by Doviak and Zrnic (138) and Atlas (3) for greater detail and further elaboration.

References

1. Hitschfeld, W., 1986: The invention of radar meteorology. *Bull. Amer. Meteor. Soc.,* **67,** 33–37.
2. Rogers, R. R., and P. L. Smith, 1996: A short history of radar meteorology. *Historical Essays on Meteorology 1919–1995,* Amer. Meteor. Soc., Boston, 57–98.
3. Atlas, D., Ed., 1990: *Radar in Meteorology.* Amer. Meteor. Soc., Boston, 806 pp.
4. Fletcher, J. O., 1990: Early developments of weather radar during World War II. *Radar in Meteorology,* D. Atlas, Ed., Amer. Meteor. Soc., Boston, 3–6.
5. Byers, H. R., and R. R. Braham, 1949: *The Thunderstorm.* U.S. Government Printing Office, 287 pp.
6. Wexler, H., 1947: Structure of hurricanes as determined by radar. *Ann. N.Y. Acad. Sci.,* **48,** 821–844.
7. Atlas, D., 1953: Device to permit radar contour mapping of rain intensity in rainstorms. U. S. Patent No. 2,656,531, Oct 20, 1953.
8. Atlas, D., 1955: Device to permit radar contour mapping of rain intensity in rainstorms. U.S. Patent reissue 24,084, Nov 1, 1955.
9. Metcalf, J. I., and K. M. Glover, 1990: A history of weather radar research in the U.S. Air Force. *Radar in Meteorology,* D. Atlas, Ed., Amer. Meteor. Soc., 32–43.
10. Austin, P. M., and S. G. Geotis, 1990: Weather radar at M.I.T. *Radar in Meteorology,* D. Atlas, Ed., Amer. Meteor. Soc., 22–31.
11. Douglas, R. H., 1990: The Stormy Weather Group (Canada). *Radar in Meteorology,* D. Atlas, Ed., Amer. Meteor. Soc., 61–68.
12. Marshall, J. S., and W. Mck. Palmer, 1948: The distribution of raindrops with size. *J. Meteor.,* **5,** 165–166.
13. Fujiwara, M., 1965: Raindrop-size distribution from individual storms. *J. Atmos. Sci.,* **22,** 585–591.
14. Battan, L. J., 1973: *Radar Observation of the Atmosphere.* University of Chicago Press, Chicago, 324 pp.
15. Atlas, D., and A. C. Chmela, 1957: Physical-synoptic variations of drop-size parameters. *Proc. Sixth Weather Radar Conf.,* Cambridge, MA, Amer. Meteor. Soc, 21–30.
16. Ulbrich, C. W., and D. Atlas, 1978: The rain parameter diagram: Methods and applications. *J. Geophys. Res.,* **83,** 1319–1325.
17. Seliga, T. A., and V. N. Bringi, 1976: Potential use of radar differential reflectivity measurements at orthogonal polarizations. *J. Appl. Meteor.,* **15,** 69–76.
18. Ulbrich, C. W., 1983: Natural variations in the analytical form of the raindrop size distribution. *J. Climate and Appl. Meteor.,* **22,** 1764–1775.
19. Atlas, D., C. W. Ulbrich, F. D. Marks Jr., E. Amitai, and C. R. Williams, 1999: Systematic variation of drop size and radar-rainfall relations. *J. Geophys. Res.,* **104, D6,** 6155–6169.
20. Zawadzki, I. I., 1984: Factors affecting the precision of radar measurement of rain. Preprints *Twenty-Second Conference on Radar Meteorology,* Zurich, Switzerland, Amer. Meteor. Soc., 251–256.
21. Plank, V. G., 1956: A meteorological study of radar angels. Geophys. Res. Paper, 52, Geophysics Research Directorate, Air Force Cambridge Research Laboratories, Bedford, Massachusetts, 117 pp.
22. Watson-Watt, R. A., A. F. Wilkins, and E. G. Bowen, 1936: The return of radio waves from the middle atmosphere. *Proc. Royal Soc.,* **161,** 81 pp.

23. Atlas, D., 1959: Radar studies of meteorological "angel" echoes. *J. Atmos. Terr. Phys.*, **15**, 262–287.

24. Appelton, E. V., and J. H. Piddington, 1938: The reflection coefficients of ionospheric regions. *Proc. Royal Soc.*, **164**, Series A, 467–476.

25. Friend, A. W., 1939: Reflection of medium and short waves in the troposphere. *Nature*, **144**, 31.

26. Crawford, A. B., 1949: Radar reflections in the lower atmosphere. *Proc. I.R.E.*, **37**, 404–405.

27. Atlas, D., 1965: Angels in focus. *Radio Science J. of Res.*, NBS/USNC-URSI, **69D**, 871–875.

28. Gossard, E. A., 1990: Radar research on the atmospheric boundary layer. *Radar in Meteorology*, D. Atlas, Ed., Amer. Meteor. Soc., 477–527.

29. Wilson, J., 1994: Boundary layer clear-air radar echoes: Origin of echoes and accuracy of derived winds. *J. Atmos. and Oceanic Tech.*, **11**, 1184–1206.

30. Atlas, D., 1960: Radar detection of the sea breeze. *J. Meteor.*, **17**, 244–258.

31. Harper, W. G., F. H. Ludlam, and P. M. Saunders, 1957: Radar echoes from cumulus clouds. *Proc. Sixth Weather Radar Conf.*, Cambridge, MA, Amer. Meteor. Soc, 267–272.

32. Plank, V. G., 1959: Convection and refractive index inhomogeneities. *J. Atmos. Terr. Phys.*, **15**, 228–247.

33. Knight, C. A., and L. J. Miller, 1998: Early radar echoes from small warm cumulus: Bragg and hydrometeor scattering. *J. Atmos. Sci.*, **55**, 2974–2992.

34. Lane, J. A., and R. W. Meadows, 1963: Simultaneous radar and refractometer soundings of the troposphere. *Nature*, **197**, 35–36.

35. Tatarski, V. I., 1961: *Wave Propagation in a Turbulent Medium*. McGraw-Hill, New York, 285 pp.

36. Smith, P. L., and R. R. Rogers, 1963: On the possibility of radar detection of clear air turbulence. *Proc. Tenth Weather Radar Conference*, Washington, D.C., Amer. Meteor. Soc, 316–322.

37. Atlas, D., K. R. Hardy, and K. Naito, 1966: Optimizing the radar detection of clear air turbulence. *J. Appl. Meteor.*, **5**, 450–460.

38. Atlas, D., K. R. Hardy, K. M. Glover, I. Katz, and T. G. Konrad, 1966: Tropopause detected by radar. *Science*, **153**, 1110–1112.

39. Ottersten, H., 1969: Atmospheric structure and radar backscattering in the clear air. *Radio. Sci.*, **4**, 1179–1193.

40. Hardy, K. H., and H. Ottersten, 1969: Radar investigations of convective patterns in the clear atmosphere. *J. Atmos. Sci.*, **26**, 666–672.

41. Hardy, K. H., and I. Katz, 1969: Probing the clear atmosphere with high power, high resolution radars. *Prof. IEEE*, **57**(4), 468–480.

42. Glover, K. M., and K. H. Hardy, 1966: Dot angels: Insects and birds. Preprints, *Twelfth Radar Meteorology Conf.*, Norman, Oklahoma, Amer. Meteor. Soc, 264–268.

43. Hardy, K. H., and K. S. Gage, 1990: The history of radar studies of the clear atmosphere. *Radar in Meteorology*, D. Atlas, Ed., Amer. Meteor. Soc, 130–142.

44. Atlas, D., M. Kerker, and W. Hitschfeld, 1953: Scattering and attenuation by nonspherical atmospheric particles. *J. Atmos. Terr. Phys.*, **3**, 108–119.

45. Seliga, T. A., R. G. Humphries, and J. I. Metcalf, 1990: Polarization diversity in radar meteorology: Early developments. *Radar in Meteorology*, D. Atlas, Ed., Amer. Meteor. Soc., 109–114.

46. Pruppacher, H. R., and R. L. Pitter, 1971: A semi-empirical determination of the shape of cloud and raindrops. *J. Atmos. Sci.*, **28**, 86–94.

47. Seliga, T. A., and V. N. Bringi, 1976: Potential use of radar differential reflectivity measurements at orthogonal polarizations for measuring precipitation. *J. Appl. Meteor.*, **15**, 69–76.

48. Rogers, R. R., 1992: John Stewart Marshall, necrology. *Bull. Amer. Meteor. Soc.,* **73,** 1464–1465.

49. Marshall, J. S., and W. Hitschfeld, 1953: Interpretation of the fluctuating echo from randomly distributed scatterers. Part I. *Canadian J. Physics,* **31,** 962–994.

50. Rogers, R. R., 1986: Walter Hitschfeld, necrology. *Bull. Amer. Meteor. Soc.,* **67,** 1159.

51. Kessler, E., and D. Atlas, 1956: Radar-synoptic analysis of hurricane "Edna." Geophysical Research Paper #50, Cambridge, Air Force Cambridge Research Laboratories, 113 pp.

52. Atlas, D., K. A. Hardy, R. Wexler, and R. Boucher, 1963: On the origin of hurricane spiral bands. *Proc. Third Technical Conf. on Hurricanes,* Mexico City, Geofisica Internatl. (Mexico), **3,** Nos. 3–4, 123–132.

53. Marks, F. D., Jr., 1990: Radar observations of tropical weather systems. *Radar in Meteorology,* D. Atlas, Ed., Amer. Meteor. Soc., 401–425.

54. Atlas, D., 1959: Radar lightning echoes and atmospherics in vertical cross section. *Recent Advances in Atmospheric Electricity,* Pergamon Press, 441–460.

55. Rodger, C. J., 1999: Red sprites, upward lightning, and VLF perturbations. *Reviews of Geophysics,* **37,** 317–336.

56. Lhermitte, R. M., and D. Atlas, 1965: Atmospheric motion coherent pulse Doppler radar systems. U.S. Patent No. 3,212,085, Oct. 12, 1965.

57. Atlas, D., 1964: Advances in radar meteorology. *Advances in Geophysics,* **10,** Academic Press, 317–478.

58. Browning, K. A., and R. Wexler, 1968: A determination of kinematic properties of a wind field using Doppler radar. *J. Appl. Meteor.,* **7,** 105–113.

59. Rogers, R. R., 1990: The early years of Doppler radar in meteorology. *Radar in Meteorology,* D. Atlas, Ed., Amer. Meteor. Soc., 122–129.

60. Armstrong, G. M., and R. J. Donaldson, 1969: Plan shear indicator for real-time Doppler radar identification of hazardous storm winds. *J. Appl. Meteor.,* **8,** 376–383.

61. Browning, K. A., and F. H. Ludlam, 1962: Airflow in convective storms. *Quart. J. Roy. Meteor. Soc.,* **88,** 117–135.

62. Donaldson, R. J., and K. H. Glover, 1980: Joint agency Doppler technology tests. AFGL-TR-80-0357, Air Force Geophysics Laboratory.

63. Kessler, E., 1990: Radar meteorology at the National Severe Storms Laboratory, 1964–1986. *Radar in Meteorology,* D. Atlas, Ed., Amer. Meteor. Soc., 44–53.

64. Ryde, J. W., 1946: The attenuation and radar echoes produced at centimeter wavelengths by various meteorological phenomena. *Meteorological Factors in Radio-Wave Propagation,* Physical Society of London, 169–189.

65. Eastwood, E., 1967: *Radar Ornithology.* Methuen, London, 277 pp.

66. Atlas, D., and S. C. Mossop, 1960: Calibration of a weather radar by using standard target. *Bull. Amer. Meteor. Soc.,* **41,** 377–382.

67. Probert-Jones, J. R., 1990: A history of radar meteorology in the United Kingdom. *Radar in Meteorology,* D. Atlas, Ed., Amer. Meteor. Soc., 54–60.

68. Browning, K. A., and G. B. Foote, 1976: Airflow and hail growth in supercell storms and some implications for hail suppression. *Quart. J. Roy. Meteor. Soc.,* **102,** 499–533.

69. Kropfli, R. A., I. Katz, T. G. Konrad, and E. B. Dobson, 1968: Simultaneous radar reflectivity measurements and refractive index spectra in the clear atmosphere. *Radio Sci.,* **3,** 991–994.

70. Dennenberg, J. N., 1971: The estimation of spectral moments. Tech. Rep. No. 23, Laboratory for Atmospheric Probing, Univ. of Chicago and Illinois Institute of Technology.

71. Atlas, D., R. C. Srivastava, R. E. Carbone, and D. H. Sargeant, 1969: Doppler crosswind relation in radio tropo-scatter beam swinging for a thin scatter layer. *J. Atmos. Sci.,* **26,** 1104–1117.

72. Atlas, D., R. C. Srivastava, and W. Marker, 1969: The influence of specular

reflections on bistatic tropospheric radio scatter from turbulent strata. *J. Atmos. Sci.,* **26,** 1118–1121.

73. Atlas, D., R. Tatehira, R. C. Srivastava, W. Marker, and R. E. Carbone, 1969: Precipitation-induced mesoscale wind perturbations in the melting layer. *Quart. J. Roy. Meteor. Soc.,* **95,** 544–560.

74. Atlas, D., and J. I. Metcalf, 1970: The birth of "CAT" and microscale turbulence. *J. Atmos. Sci.,* **27,** 903–913.

75. Gossard, E., J. H. Richter, and D. Atlas, 1970: Internal waves in the atmosphere from high-resolution radar measurements. *J. Geophys. Res.,* **75,** 3523–3536.

76. Metcalf, J. I., and D. Atlas, 1973: Microscale ordered motions and atmospheric structure associated with thin echo layers in stably stratified zones. *Boundary Layer Meteor,* **4,** 7–35.

77. Sekhon, R. S., and R. C. Srivastava, 1970: Snow-size spectra and radar reflectivity. *J. Atmos. Sci.,* **27,** 299–307.

78. Sekhon, R. S., and R. C. Srivastava, 1971: Doppler radar observations of drop-size distribution in a thunderstorm. *J. Atmos. Sci.,* **28,** 983–994.

79. Hildebrand, P. H., and R. S. Sekhon, 1972: Objective determination of the noise level in Doppler spectra. *J. Appl. Meteor.,* **13,** 808–811.

80. Hildebrand, P., R. Neitzel, R. Parsons, J. Testud, F. Baudin, and A. LeCornec, 1996: The ELDORA/ASTRAIA airborne Doppler radar: High-resolution observations from TOGA COARE. *Bull. Amer. Meteor. Soc.,* **77,** 213–232.

81. Ekpenyong, B. E., and R. C. Srivastava, 1970: Radar characteristics of the melting layer: A theoretical study. Tech. Rep. No. 16, Laboratory for Atmospheric Probing, Univ. of Chicago and Illinois Institute of Technology, 40 pp.

82. Harris, F. I., 1977: The effects of evaporation at the base of ice precipitation layers: Theory and radar observations. *J. Atmos. Sci.,* **34,** 651–672.

83. Heymsfield, A. J., 1975: Cirrus uncinus generating cells and the evolution of cirroform clouds. Part II. The structure and circulation of the cirrus uncinus generating head. *J. Atmos. Sci.,* **32,** 808–819.

84. Serafin, R. J., L. C. Peach, and D. Atlas, 1972: Radar measurements of the atmospheric turbulence structure functions. Preprints, *15th Radar Meteor. Conf.,* Champaign-Urbana, Illinois, 280–285.

85. Srivastava, R. C., and D. Atlas, 1969: Growth, motion, and concentration of precipitation particles in convective storms. *J. Atmos. Sci.,* **26,** 535–544.

86. Srivastava, R. C., 1971: Size distribution of raindrops generated by their breakup and coalescence. *J. Atmos. Sci.,* **28,** 410–415.

87. Atlas, D., R. C. Srivastava, and R. S. Sekhon, 1973: Doppler radar characteristics of precipitation at vertical incidence. *Rev. Geophys. Space Phys.,* **11,** 1–35.

88. National Academy of Sciences, 1969: Atmospheric exploration by remote probes. D. Atlas, Ed., *Report of the Panel on Remote Atmospheric Probing,* U.S. National Academy of Sciences, Vol. 1, 61 pp; Vol. 2, 698 pp.

89. Atlas, D., and C. W. Ulbrich, 1974: The physical basis for attenuation-rainfall relationships and the measurement of rainfall parameters by combined attenuation and radar methods. *J. Rech. Atmos.,* **VIII,** 275–298.

90. Knight, C. A., and P. Squires, 1982: *Hailstorms of the Central High Plains.* Colorado Associated University Press, Boulder, Colorado, 282 pp.

91. Schickedanz, P. T., and S. A. Changnon Jr., 1970: A study of crop-hail insurance records for northeastern Colorado with respect to the design of the National Hail Experiment. Illinois State Water Survey, Urbana, Illinois, Final Report to NCAR, 98 pp.

92. Foote, G. B., and C. A. Knight, Eds., 1979: Hail, a review of hail science and hail suppression. *Meteorological Monographs,* **16,** No. **38,** Amer. Meteor. Soc, 277 pp.

93. Marwitz, J. D., 1973: Hailstorms and hail suppression techniques in the U.S.S.R. *Bull. Amer. Meteor. Soc.,* **54,** 317–325.

94. Long, A. B., E. L. Crow, and A. W. Huggins, 1976: Analysis of the hailfall during 1972–1974 in the National Hail Research Experiment. *Second WMO Scientific Conf. on Weather Modification*, Boulder, Colorado. World Meteorological Organization. WMO No. **443,** World Meteorological Organization, Geneva, Switzerland, 265–272.

95. Atlas, D., 1977: The paradox of hail suppression. *Science,* **195,** 139–145.

96. Gertzman, H. S., and D. Atlas, 1977: Sampling errors in the measurement of rain and hail parameters. *J. Geophys. Res.,* **82,** 4955–4966.

97. Atlas, D., and C. W. Ulbrich, 1977: Path- and area-integrated rainfall measurement by microwave attenuation in the 1–3 cm band. *J. Appl. Meteor.,* **16,** 1322–1331.

98. Meneghini, R., J. Eckerman, and D. Atlas, 1983: Determination of rain rate from a spaceborne radar using measurements of total attenuation. *IEEE Transactions on Geoscience and Remote Sensing,* **GE-21,** 34–43.

99. Meneghini, R., and D. Atlas, 1986: Simultaneous ocean cross-section and rainfall measurements from space with a nadir-looking radar. *J. Atmos. Oceanic Technol.,* **3,** 400–413.

100. Atlas, D., C. Elachi, and W. E. Brown Jr., 1977: Precipitation mapping with an airborne synthetic aperture imaging radar. *J. Geophys. Res.,* **82,** 3445–3451.

101. Kellogg, W. W., D. Atlas, D. S. Johnson, R. J. Reed, and K. C. Spengler, 1974: Visit to the People's Republic of China: A report from the AMS delegation. *Bull. Amer. Meteor. Soc.,* **55,** 1291–1350.

102. Atlas, D., 1976: Atmospheric sciences in changing times. *Bull. Amer. Meteor. Soc.,* **57,** 573–575.

103. Atlas, D., Ed., 1976: *Atmospheric Science and Public Policy.* Amer. Meteor. Soc., 105 pp.

104. Newell, H. P., 1980: Beyond the atmosphere. SP-4211, National Aeronautics and Space Administration, Washington, DC.

105. Thomas, L., 1974: *The Lives of a Cell.* Viking Press, 192 pp.

106. Zwally, H. J., J. C. Comiso, C. L. Parkinson, W. J. Campbell, F. D. Carsey, and P. Gloersen, 1983: Antarctic Sea ice, 1973–1976: Satellite passive-microwave observations. NASA SP-459, National Aeronautics and Space Administration, Washington, DC, 206 pp.

107. Parkinson, C. L., 1983: On the development and cause of the Weddell polynya in a sea ice simulation. *J. Phys. Oceanogr.,* **13,** 501–511.

108. Atlas, D., and O. W. Thiele, 1982: Precipitation measurements from space: Workshop summary. *Bull. Amer. Meteor. Soc.,* **63,** 59–63.

109. Atlas, D., J. Eckerman, R. Meneghini, and R. K. Moore, 1981: The outlook for precipitation measurements from space. *Atmos.—Ocean,* **20,** 50–61.

110. Atlas, D., C. Ulbrich, and R. Meneghini, 1984: The multi-parameter remote measurement of rainfall. *Radio Sci.,* **19,** 3–22.

111. Chou, S. H., D. Atlas, and E. N. Yeh, 1986: Turbulence in a convective marine atmospheric boundary layer. *J. Atmos. Sci.,* **43,** 547–564.

112. Atlas, D., and T. J. Matejka, 1985: Airborne Doppler radar velocity measurement of precipitation in ocean surface reflection. *J. Geophys. Res.,* **90,** 5820–5828.

113. Meneghini, R., and D. Atlas, 1986: Simultaneous ocean cross-section and rainfall measurements from space with a nadir-looking radar. *J. Atmos. Oceanic Technol.,* **3,** 400–413.

114. Battan, L. J., 1959: *Radar Meteorology.* University of Chicago Press, Chicago, Illinois, 161 pp.

115. Fu, L., and B. Holt, 1982: SEASAT views oceans and sea ice with synthetic aperture radar. Publ. 81-120, Jet Propulstion Lab., Calif. Inst. of Technol., Pasadena, California, 200 pp.

116. Atlas, D., 1994: Footprints of storms on the sea: A view from spaceborne synthetic aperture radar. *J. Geophys. Res.,* **99, C4,** 7961–7969.

117. Atlas, D., 1994: Origin of storm SAR footprints on the sea. *Science,* **266,** 1364–1367.

118. Doneaud, A. A., S. I. Niscov, D. L. Priegnitz, and P. L. Smith, 1984: The area-time integral as an indicator for convective rain volume. *J. Appl. Meteor.,* **23,** 555–561.

119. Chiu, L. S., 1988: Rain estimation from satellites: Areal rainfall-rain area relations. *Third Conf. on Satellite Meteorology and Oceanography,* Amer. Meteor. Soc., Anaheim, CA, 363–368.

120. Calheiros, R. V., and I. I. Zawadzki, 1987: Reflectivity rain-rate relationships for radar hydrology in Brazil. *J. Climate Appl. Meteor.,* **26,** 118–132.

121. Atlas, D., D. Rosenfeld, and D. A. Short, 1990: The estimation of convective rainfall by area integrals: Part 1, theoretical and empirical basis. *J. Geophys. Res.,* **95,** 2153–2160.

122. Rosenfeld, D., D. Atlas, and D. A. Short, 1990: The estimation of convective rainfall by area integrals: Part 2, the Height-Area Rainfall Threshold (HART) method. *J. Geophys. Res.,* **95,** 2161–2176.

123. Rosenfeld, D., D. B. Wolff, and E. Amitai, 1994: The window probability matching method for rainfall measurement with radar. *J. Appl. Meteor.,* **33,** 683–693.

124. Atlas, D., and T. L. Bell, 1992: The relation of radar to cloud area-time integrals and implications for measurements from space. *Mon. Wea. Rev.,* **120,** 1997–2008.

125. Richards, F., and P. Arkin, 1981: On the relationship between satellite observed cloud cover and precipitation. *Mon. Wea. Rev.,* **109,** 1981–1093.

126. Marks, F. D., Jr., D. Atlas, and P. T. Willis, 1993: Probability matched reflectivity-rainfall relations for a hurricane from aircraft observations. *J. Appl. Meteor.,* **32,** 1134–1141.

127. Atlas, D., D. Rosenfeld, and A. R. Jameson, 1997: Evolution of radar rain measurement methods: Steps and Mis-steps. Keynote paper, Weather Radar Technology for Water Resources Management, B. Braga and O. Massambani, Eds., UNESCO Press, Montevideo, 3–67.

128. Hildebrand, P. H., W-C Lee, C. A. Walther, C. Frush, M. Randall, E. Loew, R. Neitzel, R. Parsons, J. Testud, F. Baudin, and A. LeCornec, 1996: The ELDORA/ASTRAIA airborne Doppler Weather radar: high-resolution observations from TOGA COARE. *Bull. Amer. Meteor. Soc.,* **77,** 213–232.

129. Atlas, D., C. W. Ulbrich, F. D. Marks Jr., E. Amitai, and C. R. Williams, 1999: Systematic variation of drop size and radar-rainfall relations. *J. Geophys. Res.,* **104, D6,** 6155–6169.

130. Atlas, D., and C. W. Ulbrich, 2000: An observationally based conceptual model of warm oceanic convective rain in the tropics. *J. Appl. Meteor.,* **39,** 2165–2181.

131. Atlas, D., 1987: Radar detection of hazardous small scale weather disturbances. U.S. Patent No. 4,649,388, March 10, 1987; Reissue No. 33152, Jan. 23, 1990.

132. Weber, M. E., and W. R. Moser, 1987: A preliminary assessment of thunderstorm outflow measurement with Airport Surveillance Radar. MIT Lincoln Laboratory, Project Report ATC-140, DOT/FAA/PM-86-38.

133. Weber, M. E., and T. A. Noyes, 1987: Low altitude wind shear detection with Airport Surveillance Radars: Evaluation of 1987 field measurements. MIT Lincoln Laboratory Project Report ATC-159, DOT/FAA/PS-88-10.

134. Weber, M. E., 1992: Low level altitude wind shear detection with airport surveillance. U.S. Patent No. 5,093,662, March, 1992.

135. Atlas, D., 1991: Evolution: Bats, radar, and science. *Bull. Amer. Meteor. Soc.,* **72,** 1381–1386.

136. Gilman, J. J., 1991: Research management today. *Physics Today,* **42,** 42–48.

137. Dalrymple, 1991: The importance of "small" science. *EOS, Trans. Amer. Geophys. Union,* **72,** 1–4.

138. Doviak, R. J., and D. S. Zrnic, 1993: *Doppler Radar and Weather Observations.* Academic Press, San Diego, CA 562 pp.